5th European Academic Symposium on EAF Steelmaking

IOP Conference Series: Materials Science and Engineering
Volume 1309

Oulu, Finland
5-7 June 2023

ISBN: 978-1-7138-9815-3
ISSN: 1757-8981

Printed from e-media with permission by:

Curran Associates, Inc.
57 Morehouse Lane
Red Hook, NY 12571

Institute of Physics
Dirac House, Temple Back
Bristol BS1 6BE UK

Phone: 44 1 17 929 7481
Fax: 44 1 17 920 0979

techtracking@iop.org

Additional copies of this publication are available from:

Curran Associates, Inc.
57 Morehouse Lane
Red Hook, NY 12571 USA
Phone: 845-758-0400
Fax: 845-758-2633
Email: curran@proceedings.com
Web: www.proceedings.com

TABLE OF CONTENTS

Preface to the Proceedings of the 5th European Academic Symposium on EAF Steelmaking

V-V Visuri[1] and T Echterhof[2]

[1] Process Metallurgy Research Unit, University of Oulu, PO Box 4300, University of Oulu, Oulu, Finland
[2] Department for Industrial Furnaces and Heat Engineering, RWTH Aachen University, 52074 Aachen, Germany

ville-valtteri.visuri@oulu.fi

Abstract. The European Academic Symposium on EAF Steelmaking (EASES) is a series of events targeted at researchers and doctoral candidates working with different aspects of electric arc furnace (EAF) steelmaking. Previously, four events were organised in 2015, 2016, 2018, and 2021 in Aachen, Germany. The 5th European Academic Symposium on EAF Steelmaking (EASES 2023) was organised jointly by the RWTH Aachen University (Department for Industrial Furnaces and Heat Engineering) and University of Oulu (Process Metallurgy Research Unit) and held at the University of Oulu in Oulu, Finland on 5–7 June 2023. A total of 29 presentations were held at the event, out of which 13 featured an accompanying full paper. After the symposium, the authors of the full papers were invited to submit their manuscripts to be included in the proceedings. The proceedings consist of a total of 12 peer-reviewed papers published in the *IOP Conference Series: Materials Science and Engineering*. Based on the presentations, the development of modelling and monitoring tools for process optimization, the use of biochar to replace fossil carbon for slag foaming, and the use of hydrogen as an alternative burner fuel were identified as current hot topics. In the future, the importance of the EAF process is expected to increase further due to the increasing availability of scrap and plans for ore-based production routes based on hydrogen direct reduction.

1. Background

The *European Academic Symposium on EAF Steelmaking* (EASES) emerged at the RWTH Aachen University to gather researchers working in the field of electric arc furnace steelmaking. Following a series of four successful events held in 2015, 2016, 2018, and 2021 in Aachen, Germany, EASES has established itself as a prominent platform for academia and industry experts to come together, exchange knowledge, and explore advancements in the field.

The 5th EASES marked a distinct step in the internationalization of the symposium. Organized for the first time outside Germany, the 5th EASES was held at the University of Oulu in Oulu, Finland on 5–7 June 2023. The idea was to introduce a rotating location in the upcoming events. The choice of the University of Oulu as the first location represents a milestone in the decade-long collaboration on EAF steelmaking between RWTH Aachen University and the University of Oulu. The conference was co-chaired by Ville-Valtteri Visuri (University of Oulu) together with Thomas Echterhof (RWTH Aachen University). The practicalities were handled by the event team at the University of Oulu (Riku Mattila, Ilpo Mäkelä, Henri Pauna, and Ville-Valtteri Visuri), while *Konffa* was contracted for the payment and registration system. For the first time, the symposium also had an international scientific committee, consisting of five representatives from relevant universities in Finland, Germany, Italy, and Slovenia (see **Table 1**). The scientific committee helped to promote the event and had a very positive effect on the number of presentations received and on the general interest in the event.

Table 1. Scientific committee of EASES 2023.

Member	Affiliation	Country
Dr Matti Aula	University of Oulu	Finland
Dr.-Ing. Thomas Echterhof	RWTH Aachen University	Germany
Asst. Prof. Vito Logar	University of Ljubljana	Slovenia
Assoc. Prof. Davide Mombelli	Politecnico di Milano	Italy
Assoc. Prof. Ville-Valtteri Visuri	University of Oulu	Finland

Eventually, a total of 29 presentations were held at EASES 2023, thus continuing the increasing trend of presentations in the EASES events (see **Figure 1**). In the symposium, the presentations were split into six sessions, which were further divided evenly between two days. The presentations were accompanied by 13 full papers. The EASES awards were introduced to acknowledge outstanding contributions to the symposium. To this end, papers submitted to the conference before the event were evaluated by the scientific committee. To maintain fair evaluation, committee members were prohibited from evaluating papers authored or co-authored by them. Based on the submissions, papers were ranked in two award categories: *Best Paper Award* and *Best Paper by Doctoral Researcher Award*. The winners of the awards were disclosed at the symposium dinner on June 6, 2023. Finally, to promote a high impact and visibility, it was chosen to publish peer-reviewed proceedings based on the contributions received. To this end, IOP Publishing was contacted and contracted to publish the proceedings.

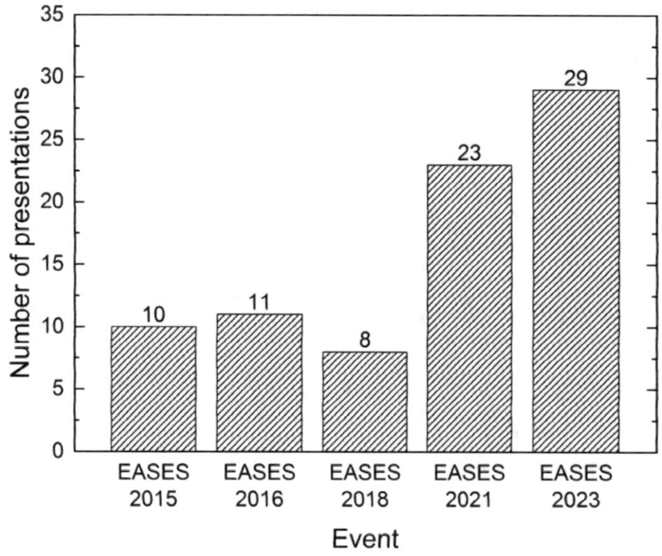

Figure 1. Evolution of the number of presentations.

2. Course of the symposium

The formal program of the EASES 2023 began on June 6, 2023, with welcoming words by the symposium co-chair Ville-Valtteri Visuri (University of Oulu). Thereafter, the technical program began with three sessions, each with five presentations (see **Table 1**). The first session (*Process control and sensors*) was chaired by Matti Aula (University of Oulu) and Henri Pauna (University of Oulu). The presentations in this session addressed topics like the implementation of new sensor technologies, computer vision or soft sensor approaches to increase the information available on the process.

Table 2. Technical presentations on June 6, 2023.

#	Title	Authors [presenter(s) underlined]
	Session I: Process control and sensors (chairs: M. Aula, H. Pauna)	
1	Electric Steelmaking Process Monitoring with Optical Emission Spectroscopy – An in-depth review	A. Rautioaho, H. Pauna, V.-V. Visuri, M. Huttula, T. Fabritius
2	An In-Situ Analysis Method in EAF and BOF Steelmaking	B. Mitas, H. Pauna, J. Feldbacher, J. Schenk
3	Soft sensor approach for continuous estimation of EAF bath temperature	A. Blažič, I. Škrjanc, V. Logar
4	Determination of the Hot Heel in an Electric Arc Furnace Using Computer Vision	A. Sharma, T. Echterhof, H. Pfeifer
5	A Decision Support System for Optimizing Electric Arc Furnace Operations using Mechanistic and Data-driven Models	S. Tomažič, V. Logar, I. Škrjanc
	Session II: Process modeling and simulation I (chairs: A. Conejo, C. Wang)	
6	A Numerical Study on the Influence of External Magnetic field on Hydrogen Electric Arc Flow	M. Al Nasser, E. Karimi-Sibaki, H. Barati, M. Wu, C. Redl, A. Ishmurzin, N. Voller, G. Hackl, A. Kharicha
7	3D Modelling of Electro Vortex Flow Inside Liquid Metal and Effect of External Magnetic Fields	M. Al Nasser, E. Karimi-Sibaki, H. Barati, M. Wu, C. Redl, A. Ishmurzin, N. Voller, G. Hackl, A. Kharicha
8	Fluid flow simulation of an EAF with bottom gas injection and coherent jets	E. Jardon, S. Fornah, A. Conejo
9	Hydrogen-Ready Energy-Efficient Technologies for EAF Steelmaking	P. Kwaschny, J. von Scheele, D. Muren
10	A Review of Thermodynamic and KineticModels for the behavior of Nitrogen in an Electric Arc Furnace	S. Nachankar, T. Echterhof, H. Pfeifer
	Session III: Process modeling and simulation II (chairs: V. Logar, P. Sulasalmi)	
11	On the differences of modelling scrap- and DRI-based EAF processes	V.-V. Visuri, I. Mäkelä
12	EAF process simulation on using hydrochar as a carburization agent	C. Wang, M. Lindvall
13	Coupled dynamic modelling of scrap melting and gas phase reactions in the EAF process	I. Mäkelä, V.-V. Visuri, M. Aula, T. Echterhof
14	Theoretical electric arc furnace model for online estimation of the unmeasured process variables	V. Logar, I. Škrjanc
15	Simulation and analysis of the EAF-based steel production process using dynamic models	D. Papamantellou, S. Chatterjee, S. Panda

The second and third sessions focused on process modeling and simulation. The second session (*Process modeling and simulation I*) was chaired by Alberto Conejo (University of Science and Technology Beijing) and Chuan Wang (Swerim). This session featured presentations on detailed 3D simulations of the electric arc and fluid flow in the EAF as well as considerations on hydrogen burners and behaviour of nitrogen. The results of the arc simulations by Mohamad Al-Nasser and co-authors highlight the highly transient behaviour of the arc and its sensitivity to external magnetic fields. As suggested in a recent review [1], by extracting information from detailed arc models, one could improve the descriptions for arc behaviour in EAF process models.

The third session (*Process modeling and simulation II*) was chaired by Vito Logar (University of Ljubljana) and Petri Sulasalmi (University of Oulu). One of the takeaways of the presentation by Ville-Valtteri Visuri and Ilpo Mäkelä was that while current process models can be applied to simulate EAF practices based on the use of direct reduced iron (DRI) and hydrogen burners, more validation with experimental data is still needed, e.g., on melting time of DRI under industrial conditions and the heat transfer efficiency of hydrogen burners in comparison to natural gas burners. On the other hand, it was also showcased that parametrised models can yield models with higher predictive performance than those based purely on mathematical descriptions for deterministic phenomena. It was recommended that overly complicated descriptions of phenomena that cannot be validated should be avoided in models intended for process control or online use.

IOP Conf. Series: Materials Science and Engineering 1309 (2024) 011001 doi:10.1088/1757-899X/1309/1/011001

The end of third session marked the end of the day's scientific part and was followed by one of the highlights of the event: the symposium dinner. It was held at Ravintola Rauhala and gave another glimpse into the academic life at the University of Oulu. The Best Paper Award was handed out for the paper "Manipulation of the leaching behavior of high-alloyed steel EAF slags through the particle size modulation" by Davide Mombelli and co-authors; the peer-reviewed version of this paper is published as Ref. [2]. The award included a free publication quota in the journal *Metals* sponsored by MDPI. The Best Paper by Doctoral Researcher Award was handed out for the paper "Application of hydrogen operated burners in the electric arc furnace" by Lilly Schüttensack and co-authors; the paper is available in peer-reviewed form in Ref. [3]. The dinner was rounded off by a performance of the band Teekkaritorvet, made up of students from the university, who performed jazz classics.

The scientific program resumed on the following day (June 7, 2023) with the remaining sessions. Chaired by Rita Kallio (University of Oulu) and Davide Mombelli (Politecnico di Milano), the fourth session focused on slags and by-products. The presentation held by Alice Petrucciani and Antonella Zaccara illustrated the use of different data-driven modelling approaches as a part of decision support systems to optimize slag reuse and recycling to reduce the environmental impact of the EAF process and slag disposal costs. A key takeaway from the presentation by Davide Mombelli was that the particle size reduction process strongly affects the leaching behavior of EAF slags by changing the nature and distribution of the main crystalline phases; finer granularity fractions had a negative influence on the leaching behaviour.

Table 3. Technical presentations on June 7, 2023.

#	Title	Authors [presenter(s) underlined]
	Session IV: Slag and by-products engineering, processing and valorisation (chairs: R. Kallio, D. Mombelli)	
16	New methodology for estimating the influence of Cu content of Ferrous Scrap materials in the Total Cost of Operation (TCO) of scrap mix in Electric Arc Furnace steelmaking	A. V. Rojo, E. Nuñez
17	Evaluating suitability of bricks to be charged to electric arc furnace	A. Abdelrahim, M. Aula, T. Fabritius
18	Laboratory Scale Evaluation of the Slag Foaming Behavior	A. Pfeiffer, K. Thiele, G. Wimmer, J. Schenk
19	Forecasting slag compositions for the optimal valorisation of electric steelworks slags	A. Petrucciani, A. Zaccara, I. Matino, M. Vannucci, V. Colla
20	Manipulation of the leaching behavior of high-alloyed steel EAF slags through the particle size modulation	D. Mombelli, G. Dall'Osto, S. Cozzo, C. Mapelli
	Session V: CO_2 emission reduction and environmental impact (chairs: T. Echterhof, S. Tomažič)	
21	Pathways for the Environmental Impact Mitigation of the Scrap Recycling Route in Italy: Possibilities and Criticalities	G. Dall'Osto, G. Villa, D. Mombelli, S. Barella, A. Gruttadauria, C. Mapelli
22	Decreasing the environmental impact of the electric steel making process through implementation of retrofitting solutions	F. Kaiser, T. Reichel, T. Echterhof, D. Mier, I. Unamuno, H. Pfeifer
23	Low impact Electric Arc Furnace: how to combine de-carbonization, energy saving and limited environmental emissions in modern electric steelmaking	A. Lanari, M. Daita
24	Biochar as a slag foaming agent in EAF – A novel experimental setup	E. Hoikkaniemi, P. Sulasalmi, V.-V. Visuri, T. Fabritius
25	Application of hydrogen operated burners in the electric arc furnace	L. Schüttensack, T. Echterhof, H. Pfeifer
26	Feasibility of Direct Carbon Fuel Cells as a Synergistic Energy Source in the Scrap Recycling Route	G. Dall'Osto, D. Mombelli, A. Pittalis, C. Mapelli
	Session VI: Fossil-free raw materials in the EAF (chairs: I. R. Souza Filho, P. Sulasalmi)	
27	GISH (Grid Integrated Steelmaking with Hydrogen)	J. Govro
28	Fundamentals of the hydrogen plasma reduction of iron ores	I. R. Souza Filho, D. Raabe, H. Springer
29	Laboratory-Scale Multi-Electrode Arc Furnace Concept for the H_2 Plasma Smelting Reduction of Iron Ore	F. Hoffelner, J. Schenk, M. Zarl

The fifth session, chaired by Thomas Echterhof (RWTH Aachen University) and Simon Tomažič (University of Ljubljana), was dedicated to the reduction of CO_2 emissions and environmental impact of EAF steelmaking. The presentation by Eetu Hoikkaniemi introduced laboratory-scale injection equipment for studying the use of biochar for slag foaming. Although further studies are needed for biochars of lower quality, the results obtained so far for high-quality biochar indicate that such biochar can, at least under laboratory conditions, perform as well as coke dust. Thomas Echterhof presented process simulations on the substitution of natural gas burners with hydrogen burners. The overall effect on the EAF process was predicted to be relatively small, affecting mainly the offgas composition. The free-flame trials with different hydrogen/natural gas ratios revealed that the addition of hydrogen stabilised the flame.

The sixth and final session focused on fossil-free raw materials and was chaired by Isnaldi R. Souza Filho (Max-Planck-Institut für Eisenforschung GmbH) and Petri Sulasalmi (University of Oulu). The session was opened by Joe Govro's presentation on the hydrogen-based production of DRI. It was suggested that in comparison to melting C-free DRI pellets, the melting of C-containing pellets benefits from CO generation, which increases the convective heat transfer. The melting time of C-free DRI pellets was thus substantially longer than that of DRI pellets containing 1.3 wt-% C. The melting time and temperature of both DRI pellet types could be predicted reasonably well using a finite element method (FEM) model executed in COMSOL. The following presentation by Isnaldi Souza Filho focused on hydrogen plasma smelting reduction (HPSR), a promising new technology that could potentially be used as a single- (without pre-reduction) or two-step (with pre-reduction) process to produce metals from oxidic materials and ores. The final presentation by Felix Hoffelner continued with the HPSR topic by introducing the ongoing work towards a laboratory-scale multi-cathode HPSR furnace at the Montanuniversität Leoben.

At the end of the conference, co-host Thomas Echterhof thanked everyone for contributing to a successful EASES 2023 by presenting, chairing a session, or just being there and participating in the lively discussions. Special thanks went to co-host Ville-Valtteri Visuri and colleagues from the University of Oulu for organising the event. Finally, it was announced that the 6th EASES is scheduled to be held in Slovenia in 2025 and will be hosted by the University of Ljubljana. A group photo was taken to commemorate the participants (see **Figure 2**).

Figure 2. A photograph of participants on the second day.

3. Proceedings

After the symposium, the authors of the full papers were invited to submit their manuscripts to be included in the proceedings. Each paper was subjected to single anonymous peer-review by two independent reviewers. After peer-review and revisions, 12 manuscripts were accepted to be published in the *IOP Conference Series: Materials Science and Engineering* (see **Table 4**). The proceedings are indexed in all the main indexing services, including *Web of Science* and *Scopus*.

Table 4. Peer-reviewed manuscripts published as part of the proceedings.

#	Title	Authors	Ref.
	Section I: Process control and sensors		
1	Electric steelmaking process monitoring with optical emission spectroscopy – An in-depth review	A. Rautioaho, H. Pauna, V.-V. Visuri, M. Huttula, T. Fabritius	[4]
2	An In-Situ Analysis Method in EAF and BOF Steelmaking	B. Mitas, H. Pauna, J. Feldbacher, J. Schenk	[5]
	Section II: Process modelling and simulation I		
3	A Numerical Study on the Influence of External Magnetic field on Hydrogen Electric Arc Flow	M. Al Nasser, E. Karimi-Sibaki, M. Wu, C. Redl, A. Ishmurzin, N. Voller, G. Hackl, A. Kharicha	[6]
4	3D Modelling of Electro Vortex Flow Inside Liquid Metal and Effect of External Magnetic Fields	M. Al Nasser, E. Karimi-Sibaki, M. Wu, C. Redl, A. Ishmurzin, N. Voller, G. Hackl, A. Kharicha	[7]
	Section III: Process modelling and simulation II		
5	Coupled dynamic modeling of scrap melting and gas phase reactions in the electric arc furnace process	I. Mäkelä, V.-V. Visuri, M. Aula, T. Echterhof	[8]
	Section IV: Slag and by-products engineering, processing and valorisation		
6	New methodology for estimating the influence of Cu content of Ferrous Scrap materials in the Total Cost of Operation (TCO) of scrap mix in Electric Arc Furnace Steelmaking	E. Nuñez, H. Priyadarshi, M. Piccoc, A. Vicente	[9]
7	Laboratory Scale Evaluation of the Slag Foaming Behavior	A. Pfeiffer, K. Thiele, G. Wimmer, J. Schenk	[10]
8	Manipulation of the leaching behaviour of high-alloyed steel EAF slags through the particle size modulation	D. Mombelli, G. Dall'Osto, S. Cozzo, C. Mapelli	[2]
	Section V: Carbon dioxide emission reduction and environmental impact		
9	Decreasing the environmental impact of the electric steel making process through implementation of furnace retrofitting solutions	F. Kaiser, T. Reichel, T. Echterhof, D. Mier, I. Unamuno, H. Pfeifer	[11]
10	Biochar as a slag foaming agent in EAF – A novel experimental setup	E. Hoikkaniemi, P. Sulasalmi, V.-V. Visuri, T. Fabritius	[12]
11	Application of hydrogen operated burners in the electric arc furnace	L. Schüttensack, A. Reinicke, T. Echterhof	[3]
	Section VI: Fossil-free raw materials in the EAF		
12	Development of a new laboratory-scale reduction facility for the H_2 plasma smelting reduction of iron ore based on a multi-electrode arc furnace concept	F. Hoffelner, M.A. Zarl, J. Schenck	[13]

4. Concluding remarks

The EAF process represents a major research topic in the domain of ferrous metallurgy. After four events held in Aachen, Germany, the 5th EASES held in Oulu, Finland, marked a significant step in the internationalisation and expansion of the symposium. The symposium featured a scientific committee consisting of five representatives from relevant universities in Finland, Germany, Italy, and Slovenia. A total of 29 presentations accompanied by 13 papers were held in the symposium. After the symposium, the authors of the full papers were invited to submit their manuscripts to be included in the proceedings Following a peer-review, 12 papers were accepted for publication in the conference proceedings.

Current hot topics identified based on the contributions in the symposium are the development of modelling and monitoring tools for process optimization, the use of biochar to replace fossil carbon for slag foaming, and hydrogen burners as alternatives for natural gas burners. This is an indication that despite the relatively low specific CO_2 emissions of the scrap-based steelmaking route, especially steelmakers based in Europe are interested in decreasing the direct and indirect CO2 emissions of scrap-based steelmaking even further.

In the future, the importance of the EAF process is expected to increase further. On one hand, the availability of scrap is projected to increase substantially in the coming decades [14], which could pave the way for increasing scrap-based steel production. On the other hand, several steelmakers are engaged in initiatives for CO_2-lean ore-based steelmaking with hydrogen-based direct reduction processes [15]; some of these concepts use natural gas alongside hydrogen. In the case of high-quality iron ore, the melting of the DRI would take place in an EAF, while a combined smelter–BOF route is envisioned for the further processing of DRI made from low-quality iron ore [15,16]. The smelter represents a new process for steelmaking and could thus become a major research topic in the future.

5. Acknowledgements

This work was conducted as part of the *Towards CO_2-lean electric arc furnace steelmaking through fundamental and data-driven mathematical modeling* (LEAF) and *Advanced Steels for a Green Planet* (AS4G) projects funded by the Research Council of Finland (project no. 356439) and Tiina and Antti Herlin Foundation respectively. The authors also wish to thank the members of the event team and scientific committee for their valuable contributions.

References

[1] Hay T, Visuri V-V, Aula M and Echterhof T 2021 A Review of Mathematical Process Models for the Electric Arc Furnace Process *Steel Res. Int.* **92** 2000395
[2] Mombelli D, Dall'Osto G, Cozzo S and Mapelli C 2024 Manipulation of the leaching behaviour of high-alloyed steel EAF slags through the particle size modulation *IOP Conf. Ser. Mater. Sci. Eng.* **1309** 012008
[3] Schüttensack L, Reinicke A and Echterhof T 2024 Application of hydrogen operated burners in the electric arc furnace *IOP Conf. Ser. Mater. Sci. Eng.* **1309** 012011
[4] Rautioaho A, Pauna H, Visuri V-V, Huttula M and Fabritius T 2024 Electric steelmaking process monitoring with optical emission spectroscopy – An in-depth review *IOP Conf. Ser. Mater. Sci. Eng.* **1309** 012001
[5] Mitas B, Pauna H, Feldbacher J and Schenk J 2024 An In-Situ Analysis Method in EAF and BOF Steelmaking *IOP Conf. Ser. Mater. Sci. Eng.* **1309** 012002
[6] Al Nasser M, Karimi-Sibaki E, Wu M, Redl C, Ishmurzin A, Voller N, Hackl G and Kharicha A 2024 A Numerical Study on the Influence of External Magnetic field on Hydrogen Electric Arc Flow *IOP Conf. Ser. Mater. Sci. Eng.* **1309** 012003
[7] Al Nasser M, Karimi-Sibaki E, Wu M, Redl C, Ishmurzin A, Voller N, Hackl G and Kharicha A 2024 3D Modelling of Electro Vortex Flow Inside Liquid Metal and Effect of External Magnetic Fields *IOP Conf. Ser. Mater. Sci. Eng.* **1309** 012004
[8] Mäkelä I and Visuri V-V 2024 Coupled dynamic modeling of scrap melting and gas phase reactions in the electric arc furnace process *IOP Conf. Ser. Mater. Sci. Eng.* **1309** 012005
[9] Nuñez E, Priyadarshi H, Piccoc M and Vicente A 2024 New methodology for estimating the influence of Cu content of Ferrous Scrap materials in the Total Cost of Operation (TCO) of scrap mix in Electric Arc Furnace Steelmaking *IOP Conf. Ser. Mater. Sci. Eng.* **1309** 012006
[10] Pfeiffer A, Thiele K, Wimmer G and Schenk J 2024 Laboratory Scale Evaluation of the Slag Foaming Behavior *IOP Conf. Ser. Mater. Sci. Eng.* **1309** 012007
[11] Kaiser F, Reichel T, Echterhof T, Mier D, Unamuno I and Pfeifer H 2024 Decreasing the environmental impact of the electric steel making process through implementation of furnace retrofitting solutions *IOP Conf. Ser. Mater. Sci. Eng.* **1309** 012009
[12] Hoikkaniemi E, Sulasalmi P, Visuri V-V and Fabritius T 2024 Biochar as a slag foaming agent in EAF – A novel experimental setup *IOP Conf. Ser. Mater. Sci. Eng.* **1309** 012010
[13] Hoffelner F, Zarl M A and Schenck J 2024 Development of a new laboratory-scale reduction facility for the H_2 plasma smelting reduction of iron ore based on a multi-electrode arc furnace concept *IOP Conf. Ser. Mater. Sci. Eng.* **1309** 012012
[14] World Steel Association 2021 Fact sheet: scrap use in the steel industry
[15] Wimmer G, Voraberger B, Kradel B and Fleischanderl A 2022 Breakthrough Pathways to Decarbonize the Steel Sector *Mitsubishi Heavy Industries Technical Review* **59** 1–7

IOP Conf. Series: Materials Science and Engineering 1309 (2024) 011001 doi:10.1088/1757-899X/1309/1/011001

[16] Voraberger B, Wimmer G, Dieguez Salgado U, Wimmer E, Pastucha K and Fleischanderl A 2022 Green LD (BOF) Steelmaking—Reduced CO_2 Emissions via Increased Scrap Rate *Metals (Basel)* **12** 466

IOP Conf. Series: Materials Science and Engineering 1309 (2024) 011002 doi:10.1088/1757-899X/1309/1/011002

Peer Review Statement

All papers published in this volume have been reviewed through processes administered by the Editors. Reviews were conducted by expert referees to the professional and scientific standards expected of a proceedings journal published by IOP Publishing.

- **Type of peer review:** Single Anonymous
- **Conference submission management system:** Morressier
- **Number of submissions received:** 12
- **Number of submissions sent for review:** 12
- **Number of submissions accepted:** 12
- **Acceptance Rate (Submissions Accepted / Submissions Received × 100):** 100
- **Average number of reviews per paper:** 2
- **Total number of reviewers involved:** 6
- **Contact person for queries:**
  Name: Ville-Valtteri Visuri
  Email: ville-valtteri.visuri@oulu.fi
  Affiliation: Oulun yliopisto

Electric steelmaking process monitoring with optical emission spectroscopy – An in-depth review

Arto Rautioaho[1], **Henri Pauna**[1], **Ville-Valtteri Visuri**[1], **Marko Huttula**[2], **and Timo Fabritius**[1]

[1] Process Metallurgy Research Unit, P.O. Box 4300, FI-90014, University of Oulu, Finland

[2] Nano and Molecular Systems Research Unit, P.O. Box 3000, FI-90014, University of Oulu, Finland

arto.rautioaho@oulu.fi

Abstract. Electric steelmaking plays an important role in various scenarios for environmentally friendly steelmaking where the processes must be energetically and economically feasible. As electric furnaces' capacities and low-grade recycled material usage can be expected to increase, optimizing the process practices and flexibility becomes paramount. The high-temperature environment of electric steelmaking sets several criteria for the implemented tools, where the equipment must withstand extreme conditions, have a low maintenance need and cost, and have the capability of real-time data acquisition and analysis. Optical emission spectroscopy (OES) has been studied in laboratory furnaces and on pilot and industrial scales to provide an *in situ* method for electric arc furnace and ladle furnace process control. Since OES is a method that measures the properties of emitted light, the applications focus on the electric arc plasma, burners' flames, and heat radiation from the molten bath. The optical spectra carry information on the composition, temperature, and status of the process. This in-depth review compiles the research and usage of OES as a process monitoring tool by focusing on electric arc plasma, burner flames, and molten bath radiation. Suggestions for further development of existing applications and potential new applications are discussed.

1. Introduction

The Intergovernmental Panel on Climate Change (IPCC) has defined the limit to global warming as 1.5 °C by 2050. The goal requires actions to reduce greenhouse gas emissions in all sectors, including industrial sectors. The steel sector is responsible for approximately 7% of all anthropogenic CO_2 emissions and steel production is predicted to grow by 25–30% by 2050 [1]. This means a huge challenge for the steel industry. In 2022, the global average of all steelmaking routes was 1.91 tons of CO2 per ton of steel [2], and this must be reduced radically. Currently, the dominant ironmaking route is based on the reduction of iron ore using coke in a blast furnace (BF) [1]. Carbon-based reduction can be described as a reaction of iron oxides and carbon, whose end-products are metallic iron and carbon dioxide [3]. Due to the high CO_2 emissions related to ironmaking process, alternative processes for iron production are needed.

The use of hydrogen as a reductant is a promising method to reduce CO_2 emissions since the end products are metallic iron and water vapor [4]. Currently, there are several research programs to replace carbon with hydrogen as a reductant [5]. One of the challenges of hydrogen reduction is that blast furnaces must be replaced with direct reduction processes, which affect the whole iron production process chain [6]. Another approach to reducing CO_2 emissions is to increase the recycling rate of steel [1]. In past decades, the use of scrap metal has increased, and a trend can be expected to continue, leading to lower quality and increased variance of recycled metal, which reduce the predictability of steelmaking processes [7]. Direct reduced iron (DRI) can be used alongside or instead of scrap in the EAF. In other words, the composition of charge material could differ in each batch, emphasizing the importance of online process control [7]. Moreover, the use of DRI substantially increases the specific energy consumption and generation of slag in the EAF [8], which stresses the importance of process optimization.

The economic aspect could be a challenge for hydrogen-reduced iron production. The BF process has matured [6], and new hydrogen-based steelmaking processes must be economically competitive with the BF-based processes. Since the final goal is CO_2-free steel production, hydrogen production must not release fossil carbon into the atmosphere. Fossil-free green hydrogen is produced through water electrolysis with green electrical energy [9] and leads to 4 to 6 times more electricity-intensive steel production than scrap metal-based steel production [10]. Better online process monitoring methods have a significant role in making new steelmaking process routes effective and flexible, leading to lower costs and decreased raw material consumption.

Due to harsh conditions in electric furnaces, there are only a few widely adopted methods to monitor the EAF process: water panel temperatures, acoustic measurements, off-gas composition analysis, and off-gas heat loss measurements. In addition, the use of accelerometers has been tested [11] but is not yet in widespread use. A significant disadvantage of these methods is that they provide only indirect information on the EAF process [12, 13]. However, the need to use varying charge materials creates a need for a direct process monitoring method so that a plant operator can make necessary changes to the process. It is expected that automation and digitalization will be increased in steelmaking in the future. Increasing computation resources creates possibilities to improve, for example, self-monitoring, self-optimization, and diagnostic [14]. The key factor of this development is reliable online data [15]. OES, as a method, aims to answer this challenge by providing additional information on process status. The method's advantage is the ability to acquire direct information about phenomena related to electric arcs and flames. In recent years, analysis methods based on OES have been studied for process control of EAF and ladle furnaces (LF). For example, it has been demonstrated that it is possible to acquire online information about the composition of LF slag [16].

This in-depth review aims to present the basic principles of the OES method and to discuss the advantages and limitations of the method compared to traditional analysis methods. The review also considers the requirements and limitations of the method. The article aims to highlight the core points of recent OES studies, from which the following conclusions can be made about the capabilities of OES:

- direct data can be acquired online from the process,
- equipment withstands the harsh industrial environment, and
- the method can be used to monitor the time development of processes.

2. OES as a method

2.1. Measurement principle

OES is based on precise analysis of electromagnetic radiation, where the applications in electric steelmaking have covered the wavelength range from ultraviolet (UV) to near-infrared (NIR). The basic idea of OES is to collect light from the light source and analyze it. The light is collected passively, so it does not require an external radiation source [15].

Electromagnetic radiation contains information on, for example, temperature, chemical elements, and compounds from the light source. Optical spectroscopy has been an essential method in astronomy for over 100 years, where objects are unreachable, and most of the information comes via electromagnetic radiation [17]. Similarities can be seen in metallurgy and astronomy: in metallurgy, conditions of heating processes may be so harsh for the measurement equipment that direct measurements are impossible, but the electromagnetic radiation from the process can be collected [18].

The principles of OES are well-known, and measurements can be done routinely in a laboratory environment, but an industrial environment sets numerous challenges for the method. For example, dust, smoke, and foaming slag may absorb or block light from an object. However, these are only challenges, not barriers, to use OES.

OES-based temperature measurement and scrap melting analysis tools are currently provided commercially for electric steelmaking [19]. Additionally, the method has many potential applications in the research phase. For example, in recent years, composition and temperature measurement in EAF and LF has been studied on a laboratory, pilot, and industrial scale with promising results. Based on previous laboratory and pilot-scale studies, Aula et al. [12] demonstrated that plasma temperature and numerous emission peaks can be measured from industrial EAF by OES. On the other hand, Pauna et al. [16] demonstrated the capability of OES to analyze the slag composition of industrial LF.

2.2. Equipment

A harsh industrial environment sets several criteria for the analysis equipment. They must be able to operate with little maintenance in a dusty environment where the temperature may vary significantly. As the principle of OES is to collect light from the object, the equipment requirements are relatively simple: optical fiber to collect and transfer light to a spectrometer, which separates light into different wavelengths, and a computer to analyze the data. Equipment can be at a safe distance from the light source; the only requirement is that the light source is at the view cone of optical fiber. Thus, optical fiber can be installed in several ways. In previous OES studies, optical fiber has been installed, for example, on the roof of the EAF [7], but measurements have also been established through casting spout of LF [20]. To protect the optical fiber, pneumatic air can be blown through the optical fiber housing to cool down the device and avoid slag splashes from reaching the fiber [7].

Another advantage of the OES method is a rapid analysis, which consists of exposure time, readout time, and data processing time. The overall timescale of one spectrum measurement depends on the visibility of the plasma and positioning of the equipment, ranging from less than 100 ms up to a couple of seconds at most [12]. Thus, OES can be used to monitor the time evolution of the processes in metallurgy.

2.2.1. Spectrometers

A spectrometer is a device that measures light intensity with respect to wavelength. In other words, the spectrometer's output data can be described as spectral energy distribution. The optical spectrometers are relatively simple devices, and their main components are slit, grating, and detector. The purpose of the slit is to narrow the light beam and control how much light reaches the detector. However, a wider slit also means poorer resolution. The purpose of grating is to reflect different wavelengths to different angles to the detector. The main parameter of grating is how many lines per mm it contains. This value controls the angular difference between two wavelengths; the more lines per mm, the greater the angular difference between two wavelengths. In practice, the finer grating splits light to a greater angle. As the detector's physical size remains the same, the detector collects light from a smaller wavelength range. This means better resolution but also less light per pixel. The detectors typically work in the same principle as the detectors of cameras, and the most used detector types in the optical area are charge-coupled device (CCD) and complementary metal–oxide–semiconductor (CMOS). As the grating splits light as a line on the detector, many spectrometer detectors consist only of one row of pixels instead of an area of pixels. However, different designs also exist. Both detector types interact with photons in the same principle: when a photon hits a pixel, the photoelectric effect releases electrons, and the number of electrons corresponds to the number of photons or light intensity [21]. The difference between CCD and CMOS detectors is how the number of electrons are transformed as a signal, resulting in differences in background noise produced by a detector. CCD detectors typically have less noise than CMOS detectors [22], which favors CCD detectors to monitor low-light objects. Spectral response (or quantum efficiency) depends on the wavelength and detector type. CCD and CMOS detectors can be used to monitor the spectrum from UV to NIR. Still, detectors are optimized for a certain wavelength range, depending on the detector model.

2.2.2. Optical fiber

The optical fiber is a "cable" that transfers light from the light source to a device that analyzes light, such as a spectrometer. The physical principle of optical fiber is based on total internal reflection. When photons hit the inner wall of fiber, they reflect inside the fiber as long as the hitting angle is below the critical angle. In practice, if the cable is intact and not extremely bent, photons can reach a critical angle only if they enter a fiber at an angle over the acceptance angle [23]. This means that optical fiber collects light from a cone-shaped area of direction where the head of the fiber is pointed. In spectroscopic use, the transmission curve is the relevant parameter of optical fiber. It is related to how different wavelengths pass the fiber. In spectroscopy, fiber material should match the wavelength range to be monitored. When the optical fiber is used with a spectrometer, it works also as a collimator.

As optical fibers and detectors respond differently to different wavelengths, a spectrometer and fiber combination should be calibrated with a standard light source, such as a stabilized tungsten-halogen light source.

3. Applications

3.1. Overview of application

In the literature, the main applications of OES in the steel industry have been temperature measurements and composition analyses. The temperature analysis is based on a heat radiation curve or monitoring emissions of certain ions or atoms. The method has been used to measure the temperature of molten baths, hot surfaces, and electric arcs. Composition analysis is based on monitoring emission peak ratios of different species from plasma or flames. For example, it can be used to evaluate the slag composition. Table 1 summarizes the recent OES studies at laboratory and industrial EAF and LF furnaces, and the following sections describe the composition and temperature analyses.

Table 1. Overview of the recent OES studies

Year	Ref.	Scale	Description	Observations
2013	[24]	Laboratory	The investigation about usability of OES for the analysis of EAF.	The work demonstrates that chemical information can be obtained from the oscillating plasma of the electric arc.
2014	[25]	Laboratory	The research about slag composition measurement in the case of fluctuating arc in EAF.	The results suggest that OES is a suitable method for determining high CrO_x and MnO content of slag despite arc fluctuation.
2014	[12]	Industrial	The research of the usability of OES for monitoring industrial EAF.	The study suggests that OES is a feasible method for industrial EAF. Plasma temperatures up to 7000 K were measured, and some chemical elements could be identified in the spectra.
2015	[13]	Industrial	The research analyzes surface conditions of slag and alkali emissions during different process periods.	The results indicate that OES can be used for monitoring scrap melting and slag surface conditions.
2019	[26]	Pilot	The investigation of the properties of electric arc by using OES combined with image analysis in EAF.	Plasma characteristics and dimensions of the electric arc were determined in pilot-scale AC EAF using OES and image analysis.
2020	[7]	Industrial	The research of the OES usability to measure the electric arc and surface temperature of the molten bath. The study considers measurements of high alloyed and carbon steel grades.	High-quality spectra were observed for high-alloyed steel grades, but foaming slag hindered visibility in the case of carbon steel grades.
2020	[16]	Industrial	The investigation of the usability of OES for composition analysis in ladle furnaces.	The study found Ca, Mg, and Mn emission peaks that can be used to evaluate CaF_2/MgO and MnO/MgO ratios in the slag. XRF analysis was used to calibrate OES results.
2021	[20]	Industrial	The investigation of the possibility of measuring CaO and CaF_2 content in the slag.	The study shows that the slag's CaO and CaF_2 content can be evaluated by OES 10 to 41 minutes before XRF analysis is finished.

3.2. Online composition analysis

Composition analyses play a significant role in the steel industry for quality assurance and process functionality. For example, slag composition has traditionally been monitored with X-ray fluorescence (XRF) analyses. The problem is that the results cannot be obtained in real-time. The most significant advantage of the OES method is that it allows acquiring information on, for example, the composition of the slag in real-time, which can make the process more efficient and improve the quality of the products [16].

Two unique features can be observed from the spectrum: continuous heat radiation curve and characteristic spectrum peaks. The physical background of characteristic peaks is related to the energy states of atoms and molecules. A high-temperature environment excites electrons bound to atoms to a higher unstable energy state, which spontaneously relaxes back to the ground state. The relaxation process generates photons whose energy corresponds to the energy difference between the excited and ground states. Each element has different energy states due to their unique structure; thus, information about composition can be obtained by analyzing characteristic peaks [12, 27].

In addition to atoms, molecules can be identified from the spectra. Electrons of the outer orbital of atoms take part in the formation of molecular bonds, affecting the energy states. In molecules, the electronic transitions occur in a more complicated and dynamic structure of states, leading to a more complicated emission peak structure. The more complex the molecule, the stronger the effect is. Characteristic peaks may disappear entirely in the case of complex molecules, making their identification difficult with the OES method [28]. However, electron excitations may also break a molecular bond. The hot environment excites the electrons constantly, from which it follows that the more complex the molecule, the more probably it is unstable in a hot environment [29]. In other words, this means that flames and plasma contain primarily ions, atoms, and relatively simple molecules, and they can be identified by OES. Figure 1 shows an example of an electric arc spectrum between two graphite electrodes without sample, showing the molecular optical emissions from cyanide (CN) and diatomic carbon (C_2) radicals. In summary, OES is a powerful method for measuring the composition of flames and plasmas [15].

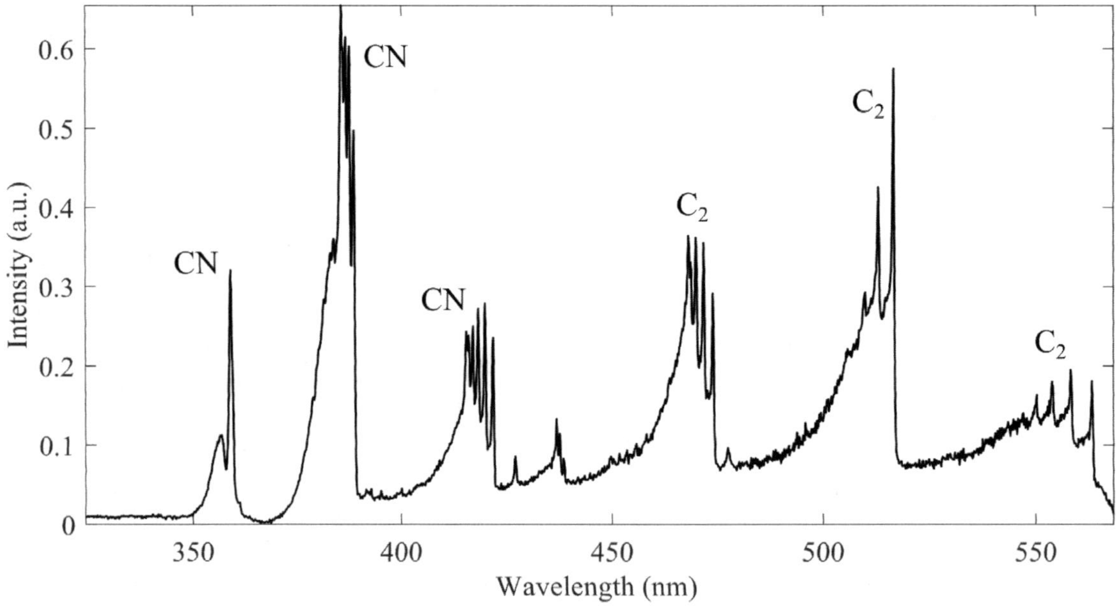

Figure 1. Optical emission spectrum from a laboratory-scale electric arc between two graphite electrodes without any samples

Composition analyses are based on the measurement of the characteristic emission peaks. The emission peaks themselves only indicate the presence of the element or molecule related to the peak but not their quantity. However, the intensity of the emission peaks gives a hint of the quantity, but intensity alone is not enough because the total intensity of a light source may vary considerably, affecting the intensity of the emission peaks. Therefore, the intensity of several emission peaks must be determined by peak fitting, for example, with Lorenz or Gauss functions [30], and the analysis is based on the relative intensities of the peaks. In some cases, intensities of emission peaks could be difficult or even impossible to measure if two or more peaks overlap. For this reason, emission peaks must be chosen carefully, case by case. Data must also be calibrated with an established analysis method, such as XRF results.

In the context of metallurgy, the applicability of OES for composition measurement has been investigated in several studies. For example, Aula et al. [25] showed that OES is the suitable method to estimate high CrO_x and MnO content on the slag, despite the arc fluctuations, and Pauna et al. [20] evaluated CaF and CaO content of slag of LF through casting spout. In this study, optical emissions of CaO, CaF, and Ca I were correlated with XRF analysis. Study shows that spectral analysis could be attributed to the CaO and CaF_2 content of the slag. The study also states that the erratic behavior of the arc must be considered in the analysis. Similarly, Pauna et al. [7] used OES to measure the optical emission from an industrial EAF furnace. The study compared the measurements of carbon steel and high alloyed steel grades. In this study, the slag foaming hindered the measurements of carbon steel but also stated that positioning and aiming of the measurement head affect the visibility. However, this study obtained promising results for high alloyed carbon steel grades: OES proved to be a potential method for online measurements of slag composition. Data could be acquired during the last 30 minutes before tapping when the charge material is completely melted. At this point, the slag composition is close to the final composition. In these studies, the data acquisition was made online, but the data was analyzed afterward.

When considering real-time data analysis and process monitoring, a balance between the number of analyzed parameters and the data processing time has to be taken into account. The OES spectra may contain several thousand emission peaks with prospects to fit tens or hundreds of emission peaks per atom or molecule. Thus, the real-time applicability may suffer with an increasing number of variables if the processing time becomes too long. On the other hand, the processing time depends on the capability of the analyzing computer, and modern computers may enable real-time analysis even under a significant processing load. On the other hand, indicators based on previous experience and research can be selected from the spectrum, which can be used to monitor the process status. For example, in Figure 2, differences in species and intensities can be seen between the beginning and later part of melting. In the simplest terms, the appearance of some species, such as slag components, in the spectrum may provide sufficient information about the process status. In addition to the aforementioned, one of the challenges of real-time analysis is the lack of calibration data and software that are suitable for real-time analysis [15]. To this end, more research is needed to implement real-time electric arc analyses.

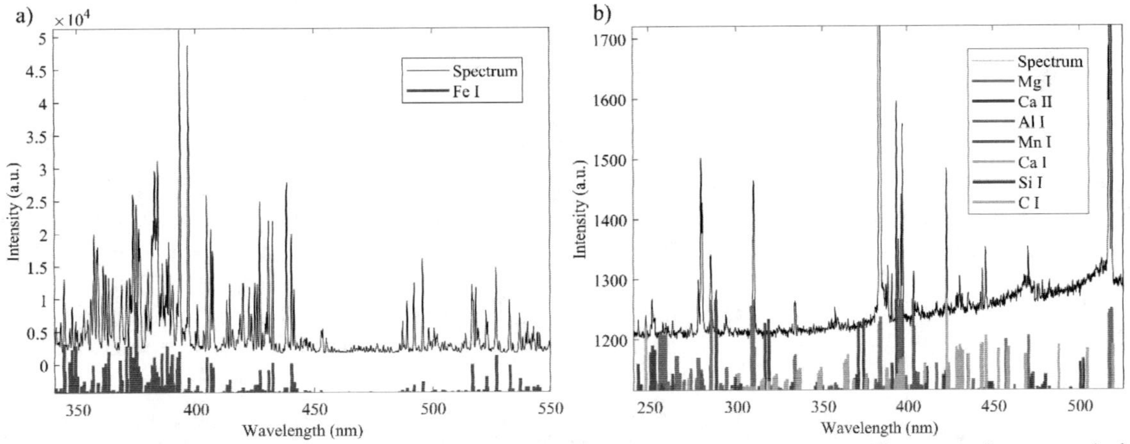

Figure 2. Example EAF spectra from a direct reduced blast furnace pellet melting experiment at a) the beginning of melting, and b) when the majority of the batch is molten. The spectra were taken in a laboratory-scale EAF at the process metallurgy research unit.

3.3. Temperature measurements

Information about temperature is one of the key factors for optimizing processes. The steel industry uses several methods to measure temperature, but the problem is that current methods cannot be used in all situations. During the EAF process, the molten bath and arc temperatures can only be estimated by using indirect methods [12]. OES is a potential method to fill this gap, enabling direct temperature measurements from the arc and the molten bath.

Temperature measurement is perhaps the most straightforward OES application to implement. There are at least two practical ways to approximate temperature with OES: comparing the measured spectrum with the thermal radiation curve and comparing the ratios of atomic emissions.

The principle of heat radiation measurement was introduced as early as the late 1800's when Wilhelm Wien introduced his displacement law. In the early 1900's, Max Planck formulated a more general form of the natural law, which describes the spectral density of electromagnetic radiation of black-body in thermal equilibrium at a certain temperature [31, 32]. A curve of radiation law can be fitted onto the measured spectrum, which gives the object's temperature. However, in the case of metallurgy, the spectra are not pure black-body radiation, but they also contain atomic and molecular emissions, which makes temperature approximation more complicated in some cases [12]. In practice, this method is useful for objects whose radiation is close to black-body radiation, for example, hot surfaces, molten baths, or flames.

The comparison of ratios of atomic emissions is based on Maxwell-Boltzmann statistics [33]. Similar to the composition analysis, this method is based on the analysis of emission peaks. However, in this case, the idea is to monitor emissions from two or more different excitation states of one specific atom or ion. The probability of which state an electron excites depends on the temperature. So, by monitoring the ratio of emissions, it is possible to get an approximation of temperature. This method is practical for measuring plasma temperatures, for example, a temperature of an electric arc [26, 34].

OES research has focused on electric furnaces so far, so the plasma temperature has mainly been measured using the method based on the Boltzmann plot. However, measurements based on the heat radiation curves have also been done. Pauna et al. [35] measured the propane oxy-fuel cutting flame temperatures. In their paper, the temperature approximated from measured spectra by using relative intensities of two wavelengths:

$$T = \left(\frac{1}{\lambda_2} - \frac{1}{\lambda_1}\right) \frac{\mathrm{hc/k}}{\ln\left[\left(\frac{I_1}{I_2}\right)\left(\frac{\lambda_1}{\lambda_2}\right)^5\right]} \tag{1}$$

where h is Planck's constant, c is the speed of light, k is the Boltzmann constant, T is the temperature, λ is the wavelength, I is the intensity, and the lower index refers to chosen wavelengths. The formula has been derived from Wien's approximation, which gives relatively accurate approximations of temperatures below 5000 K. As the method is based on black-body radiation, wavelengths must be selected carefully so that there are no atomic or molecular emission or absorption lines nearby these wavelengths. The benefit of this procedure is that the intensities are easy to obtain from the raw OES data, and the temperature approximation is simple to calculate. The temperature analysis was conducted afterward. Online analysis is possible but requires analysis software to where the data is sent in real-time.

Most of the recent OES studies have been related to electric furnaces and plasma characterization. The temperature can be derived by adding intensities of specific emission peaks to the Boltzmann equation:

$$\ln\left(\frac{\varepsilon^z \lambda_{mn}}{g_m A_{mn}}\right) = -\frac{1}{kT} E_m^z + \ln\left(\frac{hcN^z}{4\pi U^z(T)}\right) \tag{2}$$

where z is the ionization state, ε^z is the wavelength-integrated emissivity, λ_{mn} is the wavelength of emission between the upper energy state m and lower energy state n, A_{mn} is the transition probability, T is the temperature, E_m^z is the energy of the upper energy state, k is the Boltzmann constant, h is the Planck constant, c is the speed of light, N^z is the number density, and the $U^z(T)$ is the partition function [27].

The method is applicable when plasma is in thermodynamic equilibrium. Thus, McWhirter's criteria of local thermodynamic equilibrium (LTE) must be met. The criteria have been verified for laser-induced plasmas (LIP) in the range of pressure on 1.33 – 1000 hPa [27], as well as OES studies of EAF and LF furnaces [16, 26].

OES is currently the only known method that provides direct information about the temperature of an electric arc. The temperature has been studied in several research projects, and it provides information about heat radiation and gives a clue about the composition. For example, Pauna et al. [26] observed that metal vapors in the arc are connected to the relatively low plasma temperature.

4. Discussion

4.1. Requirements and limitations

Several aspects have to be considered when using OES. First of all, the method is based on the analysis of light, so the optical fiber must be placed so that the light source is in its view cone. Reflections, dust, fumes, foaming slag, scrap metal, the erratic behavior of flame or arc, etc., can disturb measurement [15, 12]. All these issues must be considered when planning the installation of optical fiber. Secondly, each OES application must be calibrated for composition analysis using an established method. For example, the composition analysis of slag must be calibrated with XRF or another method. Even if all the preparations are done correctly, the erratic behavior of plasma, flames, dust, etc., could cause temporal disturbances to the data. Therefore, the challenges of OES technology can be summarized as follows:

- poor view of the light source or unreliable line-of-sight,
- lack of calibration/reference data, and
- too many unknown phenomena in the data or process itself.

The problems mentioned above have been considered in several studies. For example, Aula et al. [25] studied OES analysis of EAF slag affected by a fluctuating arc and concluded that information on composition could be obtained despite arc fluctuation. Aula et al. [13] published another OES study about slag surface conditions and furnace atmosphere. In this study, fuming was minimized by cutting oxygen and carbon injections during measurement periods. Pauna et al. [7] studied the OES method in an industrial EAF, and they could observe high-quality spectra from a process of high alloyed steel grade. The arc was visible toward the end of the melting, which is a promising observation. The slag composition at the end of the melting is closer to the final composition, which would make the analysis more reliable. It can be concluded that these challenges can be solved, but they must be considered when using OES.

4.2. Potential other applications

Most of the OES studies in metallurgy have concentrated on electric furnaces. However, since the method is based on the analysis of light, it can measure any object that emits light. The most significant advantages of OES are its capability to acquire data rapidly and resist harsh environments with low maintenance. Especially applications that require fast analysis are the optimal use cases for OES [15].

For example, OES is the potential method for flame analysis. The light of the flame contains information about temperature and composition, which can be used to estimate radiative heat transfer and impurities. This information can be used to optimize the burning process. Pauna et al. [35] used OES together with image data to analyze an oxy-fuel-cutting process. In this case, eight spectra and images were recorded per second, which allowed linking temperature and composition to phenomena in the process, and the time development of the process could be examined.

The use of hydrogen in industrial burners is actively researched. The transition toward hydrogen is not an overnight process, and it can be assumed that hydrogen will be used together with other fuels during the transition phase [36]. OES flame analysis can facilitate the optimization of burners with different mixtures and produce information related to the hydrogen burning process. OES can be used to study heat radiation, and it can be used to study what kind of effect different ratios of hydrogen and hydrocarbon fuels have on heat radiation. This is probably essential information when new furnaces are designed and when mapping the usability of hydrogen in old furnaces.

Monitoring the time development of the process is one of the key features of OES. Possibly this could be utilized in the future by combining OES data with process automation, thus improving the efficiency of the process. The versatility of OES allows it to be used for many applications in the steel industry. For example, real-time information about the slag composition provided by OES could be fed into an automation system that automatically feeds the necessary additives. Similarly, real-time information about the temperature could be used to adjust e.g., the gas supply to the burner [15].

5. Conclusions

OES is a promising analyzing tool for the steel industry with many potential uses. As for every new measurement method, there are still challenges that need to be overcome. For example, sufficient calibration data and research on the unknown phenomena in the processes are essential for the development of OES applications Once the OES system is installed and calibrated, it can offer a low-cost and low-maintenance solution for online process monitoring. The method has already been used successfully in the following cases:

- composition analysis of slag,
- the temperature of the molten bath and plasma,
- flame analysis, and
- monitoring time development of processes.

In most cases, only the data acquisition has been made online, but the analysis has been completed afterward. The capability of OES to acquire online information can enable its use as part of the cyber-physical system. Thus, to use the method's full potential, additional research and cooperation with the industry are needed to map the OES applicability in different cases more precisely.

Acknowledgments

This work was supported by the European Union's Horizon Europe research and innovation programme under grant agreement No. 101091456 for the HyInHeat project, the Academy of Finland under the Genome of Steel grant No. 311934 and postdoctoral researcher grant No. 349402, and The Research Council of Finland Profi7 project H2FUTURE No. 352788. The work is also supported by the Strategic Research Council (SRC) within the Research Council of Finland decision 358422, JustH2Transit.

References

[1] L. Holappa, "A General Vision for Reduction of Energy Consumption and CO2 Emissions from the Steel Industry," *Metals,* vol. 10, no. 9, 2020.

[2] World Steel Association, "Sustainability Indicators 2023 report," 2023. [Online]. Available: https://worldsteel.org/steel-topics/sustainability/sustainability-indicators-2023-report/. [Accessed 16 1 2024].

[3] S. Ueda, K. Yanagiya, K. Watanabe, T. Murakami, R. Inoue and T. Ariyama, "Reaction Model and Reduction Behavior of Carbon Iron Ore Composite in Blast Furnace," *ISIJ Internationa,* vol. 49, no. 6, pp. 827-836, 2009.

[4] W. Liu, H. Zuo, J. Wang, Q. Xue, B. Ren and F. Yang, "The production and application of hydrogen in steel industry," *International Journal of Hydrogen Energy,* vol. 46, no. 17, pp. 10548-10569, 2021.

[5] F. Patisson and O. Mirgaux, "Hydrogen Ironmaking: How It Works," *Metals,* vol. 10, no. 7, 2020.

[6] R. Wang, Y. Zhao, A. Babich, D. Senk and X. Fan, "Hydrogen direct reduction (H-DR) in steel industry—An overview of challenges and opportunities," *Journal of Cleaner Production,* vol. 329, 2021.

[7] H. Pauna, M. Aula, J. Seehausen, J.-S. Klung, M. Huttula and T. Fabritius, "Optical Emission Spectroscopy as an Online Analysis Method in Industrial Electric Arc Furnaces," *steel research international,* vol. 91, no. 11, 2020.

[8] M. Kirschen, T. Hay and T. Echterhof, "Process Improvements for Direct Reduced Iron Melting in the Electric Arc Furnace with Emphasis on Slag Operation," *Processes,* vol. 9, no. 2, 2021.

[9] S. Shiva Kumar and H. Lim, "An overview of water electrolysis technologies for green hydrogen production," *Energy Reports,* vol. 8, pp. 13793-13813, 2022.

[10] N. A. Ryan, S. A. Miller, S. J. Skerlos and D. R. Cooper, "Reducing CO2 Emissions from U.S. Steel Consumption by 70% by 2050," *Environmental Science & Technology,* vol. 54, no. 22, 2020.

[11] C. Schmidt, N. Andersson, P. Ljungqvist, A. Tilliander and P. Jonsson, "Estimation of the molten state of scrap in an electric arc furnace by vibration measurements," in *Proceedings of the 6th International Congress on the Science and Technology of Steelmaking, ICS 2015*, 2015.

[12] M. Aula, A. Leppänen, J. Roininen, E.-P. Heikkinen, K. Vallo, T. Fabritius and M. Huttula, "Characterization of Process Conditions in Industrial Stainless Steelmaking Electric Arc Furnace Using Optical Emission Spectrum Measurements," *Metallurgical and Materials Transactions B,* vol. 45, 31 1 2014.

[13] M. Aula, A. Mäkinen, A. Leppänen, M. Huttula and T. Fabritius, "Optical Emission Analysis of Slag Surface Conditions and Furnace Atmosphere during Different Process Stages in Electric Arc Furnace (EAF)," *ISIJ International,* vol. 55, p. 1702–1710, 2015.

[14] T. Branca, B. Fornai, V. Colla, M. Murri, E. Streppa and A. Schröder, "Current and future aspects of the digital transformation in the European Steel Industry," *Matériaux & Techniques,* vol. 108.

[15] H. Pauna, M. Aula, T. Willms, T. Echterhof, M. Huttula and T. Fabritius, "Optical emission spectroscopy as a method to improve the process automation of electric arc furnaces and ladle furnaces," *IFAC-PapersOnLine,* vol. 55, no. 2, pp. 78-83, 2022.

[16] H. Pauna, M. Aula, J. Seehausen, J.-S. Klung, M. Huttula and T. Fabritius, "Industrial Ladle Furnace Slag Composition Analysis with Optical Emissions from the Arc," *ISIJ International,* vol. 60, no. 9, 2020.

[17] J. Hearnshaw, Astronomical spectrographs and their history, Cambridge, 2009.

[18] B. Bowman and K. Krüger, Arc Furnace Physics, 2009, p. 38.

[19] LUXMET, [Online]. Available: https://www.luxmet.fi/. [Accessed 20 1 2023].

[20] H. Pauna, M. Aula, M. Huttula and T. Fabritius, "Quantifying the CaO and CaF2 Content of Industrial Ladle Furnace Slag with Optical Emissions Measured through the Casting Spout," *Steel Research,* vol. 93, no. 5, 2021.

[21] P. Magnan, "Detection of visible photons in CCD and CMOS: A comparative view," *Nuclear Instruments and Methods in Physics Research Section A: Accelerators, Spectrometers, Detectors and Associated Equipment,,* vol. 504, no. 1-3, pp. 199-212, 2003.

[22] Y. Zhou, Y. Qu, Q. Zang, B. Wei, J. Wu, G. Peng, Y. Chang and Y. Liu, "A Low Readout Noise CMOS Pixel Based on the Skipper Technology," in *2022 23rd International Conference on Electronic Packaging Technology (ICEPT)*, 2022.

[23] ThorLabs, "Numerical Aperture and Multimode Fiber Acceptance Angle," ThorLabs, 20 1 2020. [Online]. Available: https://www.thorlabs.com/newgrouppage9.cfm?objectgroup_id=14202. [Accessed 1 11 2023].

[24] A. Mäkinen, J. T. H. Niskanen and H. Aksela, "Optical emissions from a small scale model electric arc furnace in 250-600 nm region," *Review of Scientific Instruments,* vol. 84, no. 4, 2013.

[25] M. Aula, A. Mäkinen and T. Fabritius, "Analysis of Arc Emission Spectra of Stainless Steel Electric Arc Furnace Slag Affected by Fluctuating Arc Voltage," *Applied spectroscopy,* vol. 68, no. 1, pp. 26-32, 2014.

[26] H. Pauna, T. Willms, M. Aula, T. Echterhof and M. Huttula, "Pilot-scale AC electric arc furnace plasma characterization," *Plasma Research Express,* vol. 3, no. 1, 2019.

[27] C. Aragón and J. Aguilera, "Characterization of laser induced plasmas by optical emission spectroscopy: A review of experiments and methods," *Spectrochimica Acta Part B: Atomic Spectroscopy,* vol. 63, no. 9, pp. 893-916, 2008.

[28] P. Atkins and R. Friedman, Molecular Quantum Mechanics, Fourth Edition, OXFORD UNIVERSITY PRESS, 2005.

[29] H. Al-Haj Ibrahim, "Introductory Chapter: Pyrolysis," 2020.

[30] A. J. Brown, "Spectral curve fitting for automatic hyperspectral data analysis," *IEEE Transactions on Geoscience and Remote Sensing,* vol. 44, 2006.

[31] J. Crepeau, "A Brief History of the T4 Radiation Law," in *Proceedings of the ASME Summer Heat Transfer Conference 2009, HT2009*, 2009.

[32] R. Freedman and H. young, University Physics, Pearson, 2014.

[33] A. Beiser, Concepts of Modern Physics, Sixth Edition, The McGraw-Hill companies, 2003.

[34] N. Idris, T. N. Usmawanda, K. Lahna and M. Ramli, "Temperature estimation using Boltzmann plot method of many calcium emission lines in laser plasma produced on river clamshell sample," *ournal of Physics: Conference Series,* vol. 1120, no. 1.

[35] H. Pauna, M. Aula, S. Kaukonen, M. Huttula and T. Fabritius, "Optical emission spectroscopy as a tool for process control of steelmaking burners," in *Conference: 8th International Congress on the Science and Technology of Steelmaking*, Montreal, Canada, 2022.

[36] A. Kovač, M. Paranos and D. Marciuš, "Hydrogen in energy transition: A review," *International Journal of Hydrogen Energy,* vol. 46, no. 16, 2021.

IOP Conf. Series: Materials Science and Engineering 1309 (2024) 012002 doi:10.1088/1757-899X/1309/1/012002

An In-Situ Analysis Method in EAF and BOF Steelmaking

B Mitas[1,3], H Pauna[2], J Feldbacher[1] and J Schenk[1,3]

[1] Montanuniversität, Leoben, AT
[2] University of Oulu, Oulu, FI
[3] K1-MET GmbH, Linz, AT

E-mail: bernhard@mitas.at

Abstract. An analysis was conducted to gage the feasibility of using a spectroscopic method to determine the chemical composition of the liquid metal phase in EAF and BOF steelmaking. The hypothesis that certain radiation intensities found in UV- and UV-near regions of an spectrum emitted from the hot-spot in the BOF or the impingement area of oxyfuel burners in EAF steelmaking can be used for analysing the chemical components present in the liquid metal is tested. Significant lab-scale experiment results and the conclusions regarding industrial scale implementation are presented.

1. Introduction

In order to apply big data approaches, related to industry 4.0, to metallurgical processes adequate data generation methods are required. Currently, chemical composition of the metal phase in EAF and BOF steelmaking is determined by sampling the bulk, transporting the sample via pneumatic post to the lab and analysing the sample. The largest fraction of the total time delay is attributed to the pneumatic post system and only a few samples are taken during one heat. The established sampling method is incapable of providing sufficient amounts of readings within reasonable time delays. A novel method allows the continuous and in-situ chemical analysis of the steel during the process. This enables a dynamic process control providing new insights into occurring phenomena.

2. Used materials and methods

Oxygen is blown from a 1.5 mm diameter convergent nozzle onto selected samples of iron melt containing varied amounts of phosphorous (0.00 , 0.16, 0.29, 2.29 and 24.39 at %). The total weight of the metallic sample is 150 g. Magnesium oxide is used as crucible material. A coaxial acetylene oxygen flame is used for creating conditions similar to those found in the vicinity of burners in EAF steelmaking. In the second part of the blow, the coaxial flame is switched off to determine the BOF equivalent pure oxygen impingement spectra. A spectrometer is used to analyse the light emitted at the impact area of the gas jet on the surface of the liquid metal. The experimental setup is shown in Figure 1. The spectrometer data is baseline corrected and normalized. A machine learning approach [2] determines the optimal quantity representing the inherent phosphorous content of the melt. Automatic relevance determination (ard) is used as regression method.

3. Results

A correlation for each wavelength reading of the spectrometer is made and the Pearson correlation coefficient is presented in Figure 2 and Figure 3. The first graph (Figure 2) depicts results from the acetylene oxygen impingement system while the second graph (Figure 3) shows results for the pure oxygen blowing system. The relative intensities of iron I and phosphorous I emission lines [1] and the absorption spectra of phosphorous monoxide [3] are displayed in the graph. To distinguish phosphorous from iron lines, phosphorous lines are magnified and drawn from the top downwards.

In Figure 2 there is apparent clustering at high Pearson coefficients which represents indications for clear correlations between light intensity and the phosphorous content at these wavelengths. Figure 3 suggests that there is no apparent clustering or correlations occurring in the oxygen impingement system.

A detailed overview of spectra with different phosphorous concentrations is depicted in Figure 4. The optical emissions from the OH molecule are clearly visible between 306 and 315 nm in all presented spectrums. According to Huang et al. [4], phosphorous monoxide emissions are observed between 324 and 330 nm when using acetylene flames for analysing phosphorous. As Figure 4 indicates, there is optical activity within this region with higher phosphorous content and the assumed phosphorous monoxide molecular emissions between 324 and 330 nm are clearly visible only in the spectrum with the highest amount of phosphorous. When analysing Figure 2, high correlation coefficients around 355–360 nm are observed. Emission line intensities at 325 and 357 nm are correlated with the phosphorous content in Figure 4 (b) and (c), respectively. The values are scattered with phosphorous contents below 2 at%, but there is a clear difference between these low phosphorous values and a phosphorous content of 24.39 at% where the phosphorous monoxide optical emission is clearly recognisable.

A measure for the phosphorous content of the melt is created using the spectrometer data at

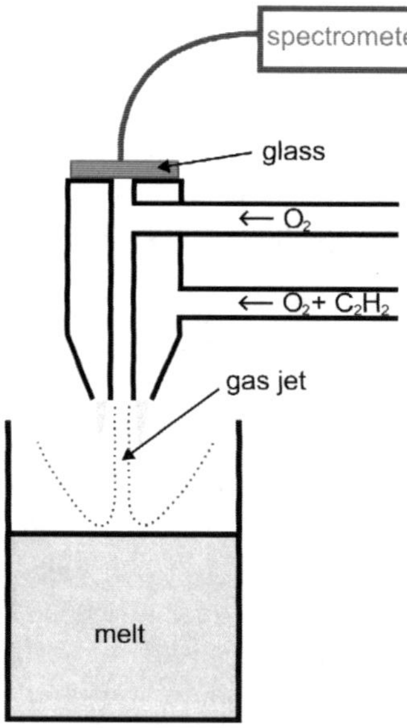

Figure 1. Experimental Setup: Oxygen is blown onto an iron melt. The emitted light at the melt surface enters a spectrometer after passing a UV-transparent glass and a fiber-optic cable. A coaxial acetylene Flame is used to influence the resulting electromagnetic spectrum. The spectrum data is analysed for predictors regarding the phosphorous content of the melt.

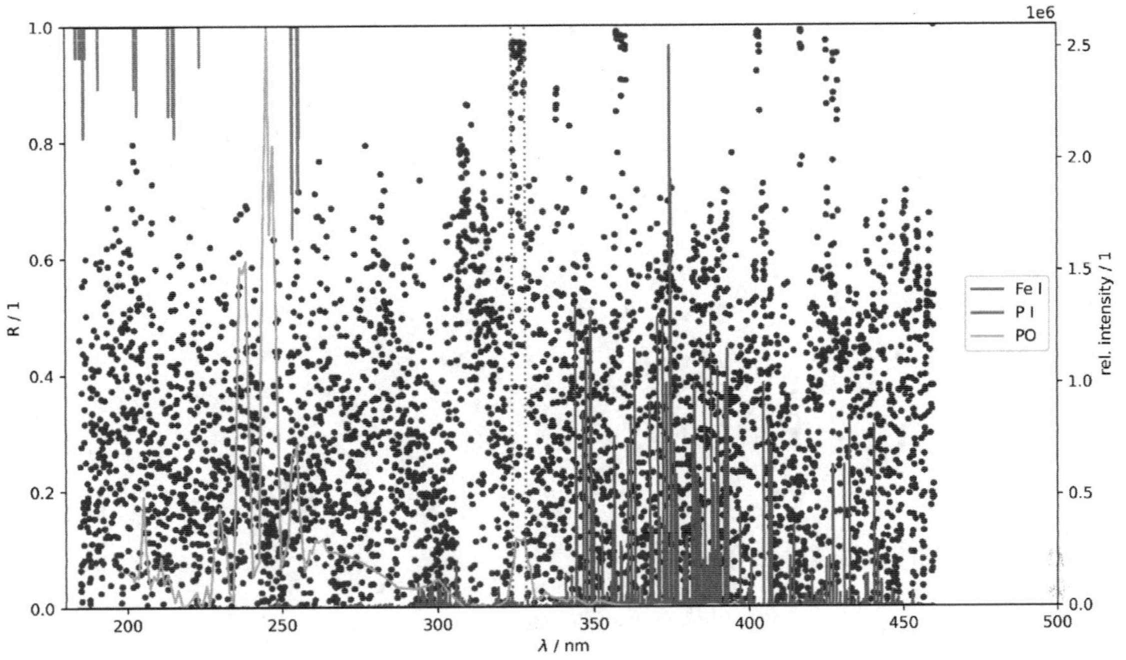

Figure 2. Oxygen jet with coaxial acetylene flame: correlation of molar fraction of phosphorus in melt with the baseline corrected relative intensity. Iron and phosphorus emission lines from the NIST database [1] and phosphorous monoxide absorption bands from Daidoji et al. [3]

various wavelengths.

4. Discussion

Daidoji et al. maintain that phosphorous as well as phosphorous monoxide exhibits emission and absorption lines in the UV region [3]. Huang et al. [4] confirmed previously found emission and absorption bands and extended the analysis regarding the molecular absorption bands of phosphorous monoxide. In this research, the Pearson correlation coefficient is greater than 95% for wavelengths between 325-328 nm (Figure 2). Daidoji et al. [3] measured phosphorous monoxide related emission and absorption phenomena within a similar region. Figure 3 does not indicate correlations in any region. Arias et al. [5] mentioned that the OH radical of a flame emits light between 306–315 nm. It can be surmised that the absence of flame radical emissions could account for the absence of measured phosphorous monoxide lines at lower wavelengths. Another reason could be the signal to noise ratio deterioration at lower wavelengths.

When blowing oxygen and acetylene onto the iron melt, some iron lines are clearly identified. The signal to noise ratio of the lines is higher for the acetylene blow. The iron lines are clearly identifiable but the phosphorous lines remain unsubstantiated. In further research, an additional analysis of the matrix effects and the emission and absorption lines would be advised. Statistical sound correlations to the phosphorous content are apparent. Subsequent research will confirm if this method can provide sufficient accuracy in measuring phosphorous contents between 0 - 1000ppm.

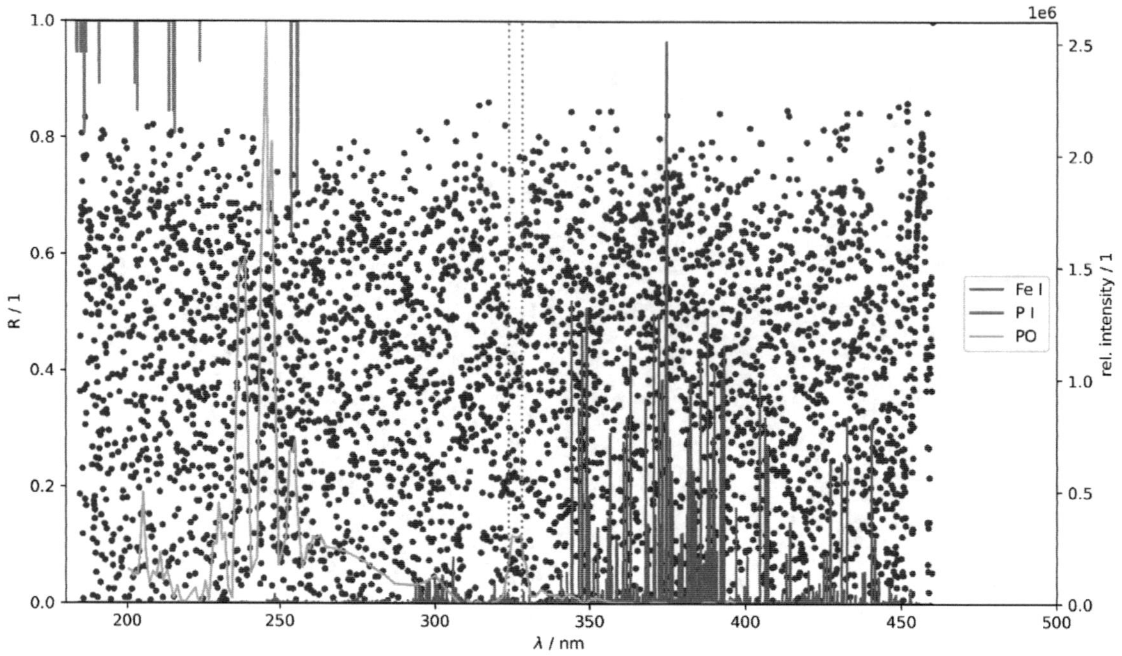

Figure 3. Oxygen jet without acetylene flame: correlation of molar fraction of phosphorus in melt with the baseline corrected relative intensity. iron and phosphorus emission lines from the NIST database [1] and phosphorous monoxide absorption spectra from Daidoji et al. [3]

5. Conclusion

Current research demonstrates how challenging it is to glean accurate findings for in-situ observations in smelting operations. Using a spectrometer, the light entering a blowing system, e.g. an oxyfuel burner or similar aggregate in EAF steelmaking or the oxygen lance or additional blowing systems in BOF steelmaking, can be used to determine important species concentration in the melt during a heat. In this research the preliminary findings of a larger study regarding in-situ phosphorous determination depict that the phosphorous content directly correlates with wavelengths between 325 and 328 nm. These correlations in the mentioned wavelength region correspond to β-phosphorous monoxide emission bands, while no α-phosphorous monoxide emission bands are characterized in this work. Machine learning algorithms are used to create predictors for the phosphorous content in the melt. This is achieved by using the automatic relevance determination method. The system predicts the phosphorous content with an average error of 370 ppm in the range of 0–10 wt% for a ten-fold cross validation. This research indicates the possibility of determining the content of various species via their oxide emission and absorption bands. Copper contents of 1 wt% could be clearly distinguished from copper free melts. Further optimisation of this method for the steelmaking processes should be encouraged and would prove beneficial.

6. Acknowledgements

The authors appreciatively acknowledge the funding support of K1-MET GmbH, a metallurgical competence center. The research program of the K1-MET competence center is supported by COMET (Competence Center for Excellent Technologies), the Austrian program for competence

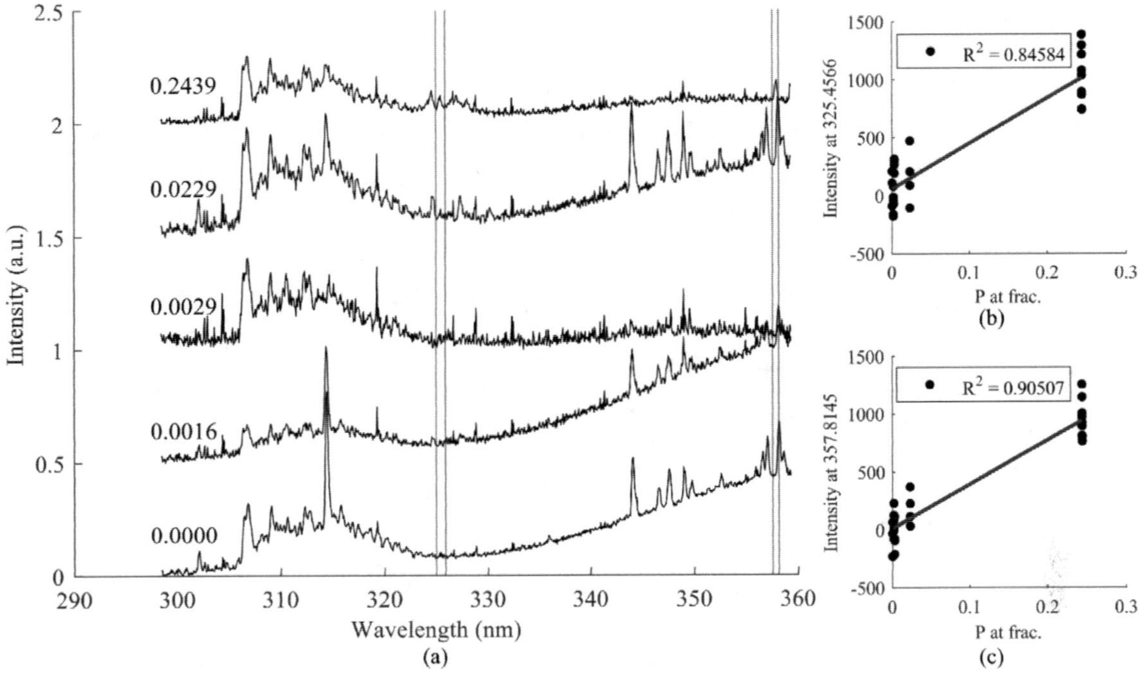

Figure 4. (a) Five spectra with atomic fractions of phosphorous written on top of each spectrum, (b) 325 nm line intensity dependence on the phosphorous content and (c) 357 nm line intensity dependence on the phosphorous content. The spectrum intensities are normalized and an off-set is applied for better readability. The positions of the (b) and (c) lines are drawn vertical in (a).

centers. COMET is funded by the Federal Ministry for Climate Action, Environment, Energy, Mobility, Innovation, and Technology, the Federal Ministry for Labour and Economy, the Federal States of Upper Austria, Tyrol, and Styria as well as the Styrian Business Promotion Agency (SFG) and the Standortagentur Tyrol. Furthermore, Upper Austrian Research continuously supports K1-MET. Besides the public funding from COMET, the current research work of K1-MET is partially financed by the participating scientific partner Montanuniversitaet Leoben and the industrial partners Primetals Technologies Austria GmbH, RHI Magnesita GmbH, and voestalpine Stahl GmbH.

The Academy of Finland is acknowledged for the postdoctoral researcher grant No. 349402.

References

[1] Kramida A, Ralchenko Y, Reader J, NIST ASD Team 2022 NIST Atomic Spectra Database (version 5.10) *NIST*
[2] Pedregosa F, Varoquaux G, Gramfort A, Michel V, Thirion B, Grisel O, Blondel M, Prettenhofer P, Weiss R, Dubourg V, Vanderplas J, Passos A, Cournapeau D, Brucher M, Perrot M, Duchesnay E 2011 *Journal of Machine Learning Research* **12** 2825 – 2830
[3] Daidoji H, Akai Y, Honma A 1979 *Bunko Kenkyu* **28** 77 – 83
[4] Huang M D, Becker-Ross H, Florek S, Heitmann U, Okruss M 2006 *J. Anal. At. Spectrom.* **21** 338 – 345
[5] Arias L, Torres S, Sbarbaro D, Farias O 2008 *Applied optics* **47** 5541 – 5549

IOP Conf. Series: Materials Science and Engineering 1309 (2024) 012003 doi:10.1088/1757-899X/1309/1/012003

A Numerical Study on the Influence of External Magnetic field on Hydrogen Electric Arc Flow

M Al Nasser[1], E karimi-Sibaki[1], M Wu[2], C Redl[3], A Ishmurzind[4], N Voller[5], G Hackl[4], A Kharicha[1]

[1]Christian-Doppler Lab for Metallurgical Applications of Magnetohydrodynamics, University of Leoben, Franz-Josef-Str. 18, 8700 Leoben, Austria
[2]Chair of Simulation and Modeling of Metallurgical Processes, University of Leoben, Franz-Josef-Str. 18, 8700 Leoben, Austria
[3] INTECO melting and casting technologies GmbH, Wienerstrasse 25, 8600 Bruck a.d. Mur, Austria
[4]RHI Magnesita GmbH, Technolgy Centre, Magnesitstrasse 2, 8700, Leoben, Austria
[5]RHI Magnesita GmbH, Kranichberggasse 6, 1120, Vienna, Austria

E-mail: abdellah.kharicha@unileoben.ac.at

Abstract. We present a 3D numerical model capable of simulating hydrogen electric arc. The numerical model couples the electromagnetic field with hydrogen plasma arc dynamics (flow and heat transfer). Electric arc operates under the conditions of atmospheric pressure using direct current (DC) in the presence of external magnetic field. It is found that arc instabilities emerge spontaneously near the cathode before propagating downwards along the arc jet. Also, spontaneous splitting into several arcs is observed before merging again into one arc. The horizontal magnetic field induces arc skewing and directional flow perpendicular to the applied magnetic field. However, the axial magnetic field induces a rotational flow and arc swirling.

1. Introduction

Several new technologies aim to decrease the carbon emission in metallurgy. One of the promising new techniques is green hydrogen technologies that rely on hydrogen instead of carbon in iron reduction. The method ultimately aims to reach a zero-carbon fingerprint in steel production. Steel produced using electric arc furnaces (EAF) exceeded 29% of total steel production in 2019. The production of steel by EAF is expected to reach 57 % by the year 2050. From this predicted share, 8% would be produced using Hydrogen gas[1]. EAF furnace produces 77 % less CO_2 than a similar size blast furnace[2]. EAF operates by transforming the electrical energy to heat by means of high-intensity electric arcs. High-intensity electric arcs operating at atmospheric pressure are coupled with extreme operating conditions like high temperatures, intense radiation, and powerful jets of plasma emitted from the cathode spot. The aforementioned conditions impose considerable difficulties in performing experimental measurements. Instead, numerical simulations of the arc offer a useful tool to understand the arc behavior without getting exposed to harsh experimental conditions.

Ushio et. Al. [3] performed one of the first simulations for the electric arc and predicted flow, thermal, and magnetic fields while assuming predefined current distribution. Szekely et al [4] further developed the model to predict the heat transfer mechanisms from arc to metal bath. Bowman [5] performed experimental measurements for a DC arc of 2160 A and measured the velocity distribution of the arc jet. The recorded velocity was up to 1500 m/s for an arc gap of 0.07 m. Jordan et al [6] measured the electrical properties and cathode spot radius of a free-burning DC arc of 10 kA. The study predicted a constant current density of 4.4×10^7 A/m^2. Bowman estimated the cathode current density to be around 3.5×10^7 A/m^2 [7]. Both estimations are still widely adopted by simulations for high-power DC arcs with Local Thermal equilibrium (LTE). The thermophysical and electrical properties of plasma for several gases including air were computed at high temperatures and generally adopted for high DC arc simulation[8,9].

Ramirez [10] simulated the arc and liquid metal bath in a DC-EAF. Cathode current density variation, arc jet velocity, the effect of compressibility, and interaction of arc with the metal bath was studied in his work. The study concluded that the metal bath does not affect the arc thermal fields so the arc can be modelled separately. Jones et al.[11] performed experimental measurements on a pilot DC EAF for currents between 2 and 7 kA. He observed extremely chaotic and spontaneous behaviour of the arc featuring several distinct structures. The different structures included diffusive arc, concentrated and violently splashing arc, pretzel-shaped arc, and occasionally arc splitting. A direct relation between arc thrust and the square of the applied current was concluded. Ramirez et al. [12] analysed the arc performance under the influence of different gases including atmospheric air, oxygen, argon, and several other gases. The study concluded that atmospheric air is the optimal medium to transfer the heat of the arc to the metal bath while Nitrogen is the least effective medium. Reynolds et al.[13,14] developed a numerical model to simulate the behaviour of DC arc plasma Single-phase arcs behaviour in addition to the arc impingement effect on the slag layer under were simulated. The model predicted a simultaneous transition between the steady and transient behaviour of the arc. However, the arc thermophysical properties were temperature independent which is far from reality. Wang et al. [15] simulated a single-phase 2D axisymmetric arc in a confined area to reproduce the effect of a submerged arc inside an EAF. Despite most previous studies being performed using 2D axisymmetric geometries, the simulations could predict very useful results[16]. The main drawback is that 2D axisymmetric is based on perfect symmetry in the azimuthal direction. On the other hand, 3D modelling enables capturing the flow instabilities. We present a numerical model capable of simulating arcs in 3D. The numerical model couples the electromagnetic field with plasma arc dynamics (flow and heat transfer). The arc dynamics and arc interaction with external magnetic fields are analyzed. The model aims to improve the understanding of arc behavior, predict and prevent failures that could occur during operation.

2. Materials and methods

In this study, the geometry consists of a cylinder full of hydrogen gas. The electric current flows between two solid electrodes in contact with the cylinder's horizontal surfaces. The top surface is in direct contact with the cathode electrode while the bottom resembles the anode surface. The lateral side of the cylinder and the top except the cathode electrode are electrically insulated. Figure 1 shows the configuration of the domain. The cylinder has a height of 25 cm (arc gap), and a radius of 25 cm. The cathode spot radius is equal to 0.6 cm. The current density at the cathode spot is assumed to be constant[6]. The temperature of the cathode electrode is set to be constant (4000 K). The cathode spot is assumed to be fixed in the center of the cathode electrode. The bottom temperature is equal to 1800 K and all other boundaries have a temperature of 300 K. Simulations are operated under local thermal equilibrium (LTE) condition. LTE enables us to relate the thermophysical and electrical properties solely to temperature, thus all hydrogen gas properties are temperature-dependent. In the current study we only assume a single-phase model where the arc strikes a flat surface. The slag/liquid metal surface is assumed to be flat. this assumption allows to model the arc separately. The flow, temperature, and electromagnetic equations were solved[17]. The LES model is used to model turbulence[18]. The applied current is 5 kA and a fixed cathode spot size is adopted. The density is incompressible temperature dependent. A fine mesh is used involving 2 million elements. The time step is chosen to be variable such that the courant number does not exceed 1.

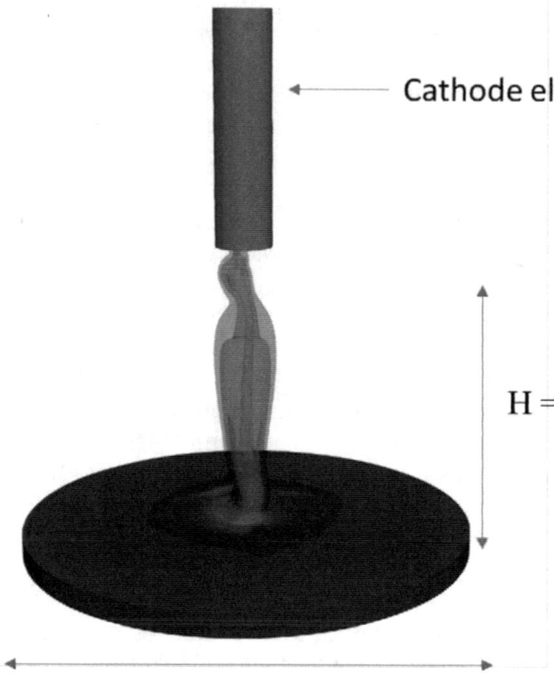

Figure 1. schematic representation of the computational domain.

3. Results

3.1. No external magnetic field:

The simulations were performed in a transient manner in the absence of external magnetic field. Figure 2 shows the arc structure and current path snapshots at different times. The vector field shows current path and direction. The transparent white surface is the isothermal surface of 10 000 K. Electrical conductivity is temperature dependent and 10000 K is considered the border line where arc exists at T>10000 K. The arc initially develops the well-known bell shape structure at time 0.00032 s. Small instabilities start at the cathode spots at t = 0.0005 s. The current entering to the cathode from the arc region interacts with the self-induced magnetic field and induces a Lorentz force that creates a powerful jet downwards[17]. Small instability develops rapidly and travels downwards along the current path. In only half a millisecond the arc structure is shifted to helical structure with kink instabilities in the arc (t = 0.001 s)[19]. These instabilities rapidly result in arc splitting. At t = 0.00132 s, two arcs of same origin can be observed spontaneously striking at two different positions on the anode surface. The current density distribution is shown on anode surface (figure 2).

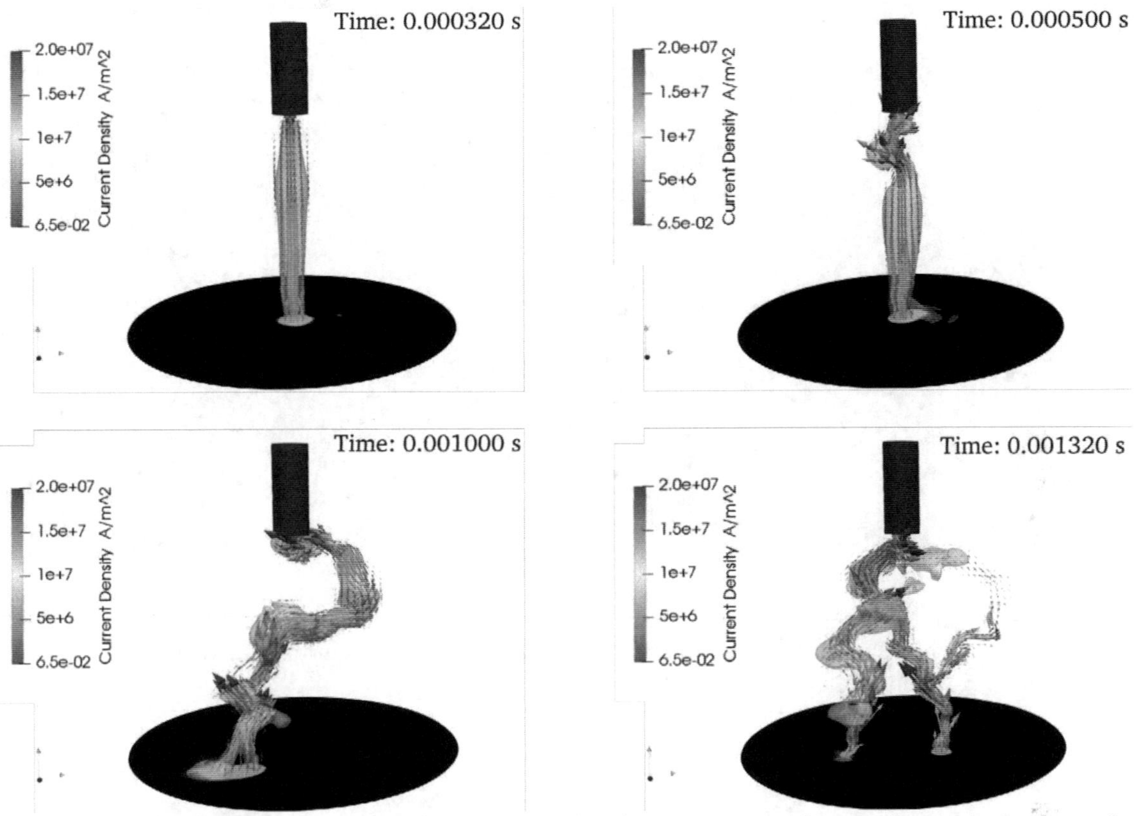

Figure 2. Snapshots at different times for current path. The transparent white surface is the isothermal surface of 10 000 K.

The aforementioned results are compared to an arc simulation in 2D axisymmetric model. Figure 3 shows a simulation for a 5-kA hydrogen arc with same arc gap (25 cm). The white isoline shows the 10000 K temperature contour. The left shows the current density and path while the right side shows velocity field and magnitude. The arc is symmetrical and centred along the axis of the domain. Instabilities appear near the cathode travel along the arc, however the simulation predicts fixed and symmetric arc around the center. The concept of symmetry in 2D falls short to predict complex flows in the arc. The stable arc in 2D model leads to relatively stable and concentrated Lorentz force at the cathode spot. This generates a powerful jet of plasma downwards. The velocity of the arc predicted by the 2D model reaches up to 10 000 m/s. On the other hand, the 3D model is able to predict more complex arc structures and is not bounded in azimuthal direction. The arc is not constricted to the center and spontaneously moves in a chaotic motion. Figure 4 shows flow distribution at two different snapshots for 3D modelling of the arc. The velocity in 3D varies along the arc and with time. The initial bell-shaped arc leads to velocity up to 4000 m/s. The instability shown in figure 2 changes the arc structure and in only 1 millisecond the flow structure changes and maximum velocity drops to 2800 m/s.

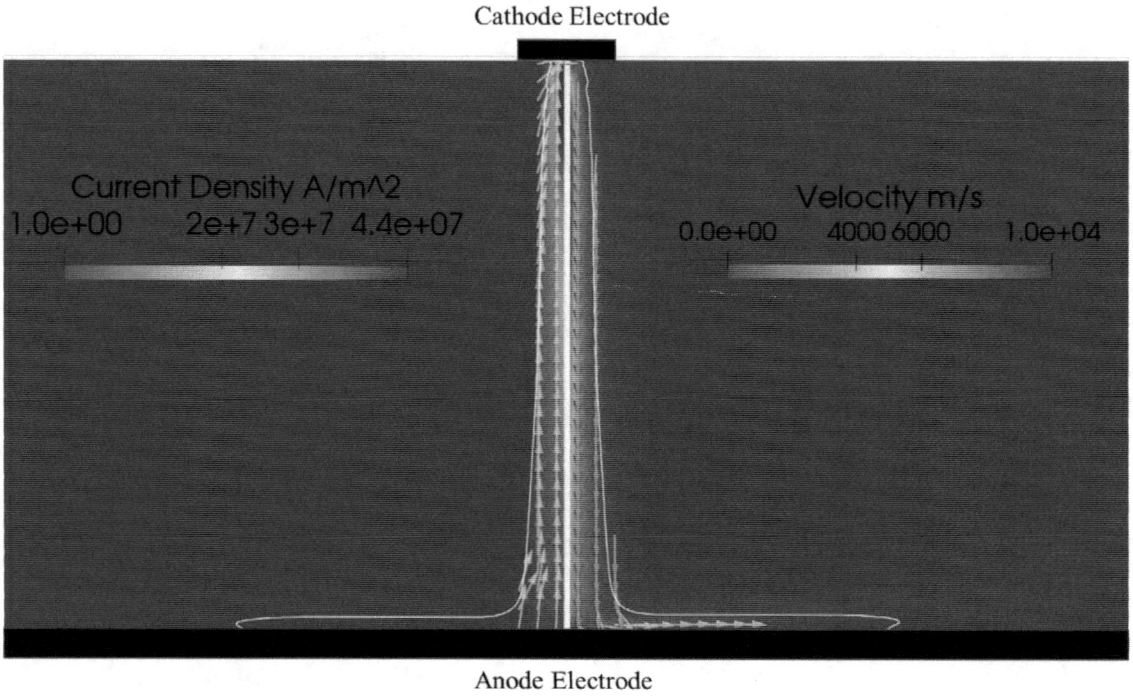

Figure 3. Arc structure in 2D axisymmetric model. The left sides show the current distribution and the right sides show the velocity distribution. The white contour is the isothermal line of 10 000 K. The vectors of both current and velocity are shown.

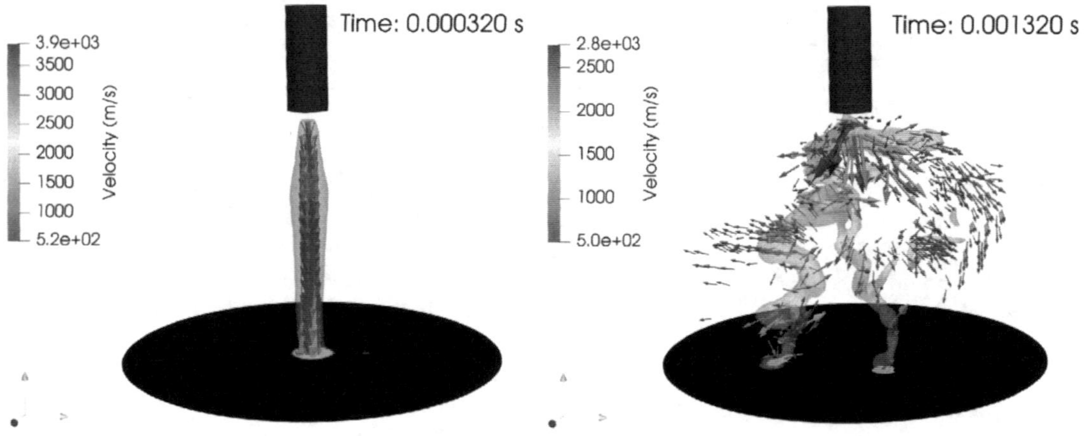

Figure 4. Velocity distribution in arc at different times. The transparent white surface is the isothermal surface of 10 000 K.

3.2. Effect of external magnetic field:

Here, effects of imposed external magnetic field on the arc dynamics and structure of the arc are investigated. It is assumed that there is a fixed external magnetic field in the domain. Figure 5. (a) shows the arc behaviour in presence of upward axial magnetic field of 0.01 T uniformly distributed in the domain. The imposed magnetic field interacts with the current in the domain and induces additional component of Lorentz force which is proportional to the current density. The arc tends to acquire a rotational behaviour in form of helical structures and deviates from the center (figure 5 a). The kink instabilities normally produce concentrated current channels. The concentrated channels of current interact with external magnetic field and produce several local helical structures along the arc path. Figure 5 (b) shows arc behaviour with a uniform horizontal magnetic field (0.01 T). The arc deviates away from the center to become unidirectional. The arc is skewed in a 90-degree angle from the imposed magnetic field. This is caused by the interaction of the horizontal magnetic field with the current at the level of the cathode spot and along the arc that creates a horizontal force affecting the arc.

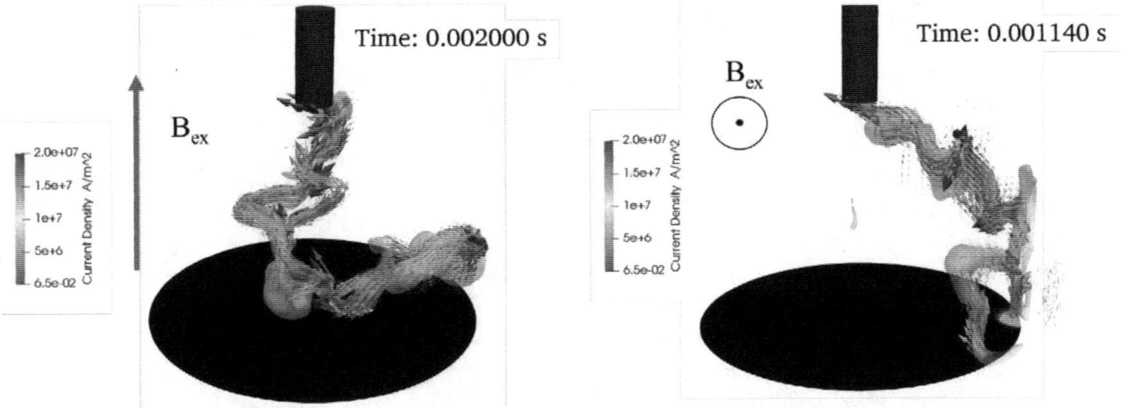

Figure 5. Arc structure in the presence of (a) axial magnetic field and (b) horizontal magnetic field. The transparent white surface is the isothermal surface of 10 000 K. Bex is the external magnetic field applied.

4. Conclusions

A magnetohydrodynamic 3D model is suggested to study the instability and behavior of a high intensity hydrogen plasma arc. The model is able to predict the transient behavior of the arc and capture several interesting behaviors such as spontaneous instabilities emerging from the cathode spot, well-known kink instabilities, and arc simultaneous splitting into several arcs. The presence of axial magnetic field alters the arc dynamics and induces rotational flow. Additionally, the presence of horizontal external magnetic field leads to horizontal directional flow perpendicular to the magnetic field direction.

Further improvements can be tackled in future to improve the presented model. These advancements include a variable size and moving cathode spot, accounting for the bore down of cathode electrode and adding a slag or liquid metal layer beneath the arc.

5. Acknowledgments

The authors acknowledge the financial support by the Austrian Federal Ministry of Economy, Family and Youth and the National Foundation for Research, Technology and Development within the framework of the Christian Doppler Laboratory for Metallurgical Applications of Magnetohydrodynamics.

References

[1] Iea 2022 Global crude steel production by process route and scenario, 2019-2050.
[2] Odenthal H J, Kemminger A, Krause F, Sankowski L, Uebber N and Vogl N 2018 Review on Modeling and Simulation of the Electric Arc Furnace (EAF) *Steel Res. Int.* **89** 1–36
[3] Ushio M and Matsuda F 1982 Mathematical Modelling of Flow Field and Heat Transfer of Welding Arc. *Int. Symp. Japan Weld. Soc.* **1** 49–54
[4] Szekely J, McKelliget J and Choudhary M 1983 HEAT-TRANSFER FLUID FLOW AND

BATH CIRCULATION IN ELECTRIC-ARC FURNACES AND dc PLASMA FURNACES. *Ironmak. Steelmak.* **10** 169–79

[5] Bowman B 1972 Measurements of plasma velocity distributions in free-burning DC arcs up to 2160 a *J. Phys. D. Appl. Phys.* **5** 1422–32

[6] Jordan G R, Bowman B and Wakelam D 1970 Electrical and photographic measurements of high-power arcs *J. Phys. D. Appl. Phys.* **3** 1089–99

[7] Bowman B 1994 Properties of Arcs in DC Furnaces *1994 Electric Furnace Conference Proceedings* pp 111–20

[8] Fauchais P and Vardelle A 1997 *Thermal plasmas* vol 25 (Springer Science & Business Media)

[9] Murphy A B 1995 Transport coefficients of air, argon-air, nitrogen-air, and oxygen-air plasmas *Plasma Chem. Plasma Process.* **15** 279–307

[10] Ramírez-argáez M A 2003 *Mathematical Modeling of DC Electric Arc Furnace Operations*

[11] Jones R T, Reynolds Q G and Alport M J 2002 DC arc photography and modelling *Miner. Eng.* **15** 985–91

[12] Ramírez-Argáez M A, González-Rivera C and Trápaga G 2009 Mathematical modeling of high intensity electric arcs burning in different atmospheres *ISIJ Int.* **49** 796–803

[13] Reynolds Q G, Jones R T and Reddy B D 2010 Mathematical and computational modelling of the dynamic behaviour of direct current plasma arcs *Proc. 12th Int. Ferroalloys Congr. Sustain. Futur.* **110** 789–801

[14] Reynolds Q G 2017 Computational Modeling of Arc–Slag Interaction in DC Furnaces *Jom* **69** 351–7

[15] Wang Z, Wang N, Li T and Cao Y 2012 3D numerical analysis of the arc plasma behavior in a submerged DC electric arc furnace for the production of fused MgO *Plasma Sci. Technol.* **14** 321–6

[16] Barati H, Kharicha A, Al-Nasser M, Kreuzer D, Hackl G, Gruber M, Ishmurzin A, Redl C, Teplyakov I O, Wu M and Ludwig A 2022 MHD instability at the cathode spot as the origin of the vortex formation in high-intensity plasma arcs *J. Phys. Commun.* **6**

[17] Al-Nasser M, Kharicha A, Barati H, Pichler C, Hackl G, Gruber M, Ishmurzin A, Redl C, Wu M and Ludwig A 2021 Toward a Simplified Arc Impingement Model in a Direct-Current Electric Arc Furnace *Metals (Basel).* **11** 1482

[18] Nicoud F and Ducros F 1999 Subgrid-scale stress modelling based on the square of the velocity *Flow Meas. Instrum.* **62** 183–200

[19] Bowman B and Krüger K 2009 *Arc Furnace Physics* (Düsseldorf: Verlag Stahleisen)

IOP Conf. Series: Materials Science and Engineering 1309 (2024) 012004 doi:10.1088/1757-899X/1309/1/012004

3D Modelling of Electro Vortex Flow Inside Liquid Metal and Effect of External Magnetic Fields

M Al Nasser[1], E karimi-Sibaki[1], M Wu[2], C Redl[3], A Ishmurzin[4], N Voller[5], G Hackl[4], A Kharicha[1]

[1]Christian-Doppler Lab for Metallurgical Applications of Magnetohydrodynamics, University of Leoben, Franz-Josef-Str. 18, 8700 Leoben, Austria
[2]Chair of Simulation and Modeling of Metallurgical Processes, University of Leoben, Franz-Josef-Str. 18, 8700 Leoben, Austria
[3] INTECO melting and casting technologies GmbH, Wienerstrasse 25, 8600 Bruck a.d. Mur, Austria
[4]RHI Magnesita GmbH, Technolgy Centre, Magnesitstrasse 2, 8700, Leoben, Austria
[5]RHI Magnesita GmbH, Kranichberggasse 6, 1120, Vienna, Austria

E-mail: abdellah.kharicha@unileoben.ac.at

Abstract. Electrovortex flow emerges when the current inside conducting liquids interacts with its self-induced magnetic field. The flow structure and strength of the flow are dictated by the current value and the presence of an external magnetic field. We present here 3D simulations for the electrovortex flow inside a liquid metal cylinder. The results presented reveal a typical electrovortex structure for low currents. Higher currents induce turbulence inside the electrovortex flow without any presence of an external magnetic field. In presence of Earth's magnetic field, the flow structure is affected significantly. Cyclone, tornado, and rope tornado are observed inside the domain due to the earth's magnetic field depending on the ratio of current applied and the earth's magnetic field.

1. Introduction

Magnetohydrodynamic (MHD) phenomena occur in many industrial processes when interactions between magnetic field and flow inside conducting liquids is present. Electrovortex flow (EVF) is an MHD effect that occurs when a current is applied inside a conducting liquid like liquid metal. The current density converges or diverges at the level of electrodes and interacts with the self-induced magnetic field to create a Lorentz force. The Lorentz force creates the flow known as EVF. Several industrial processes contain EVF such as electroslag re-melting, electric arc furnaces, aluminium reduction cell, and liquid metal batteries (1). Shercliff (2) was one of the first to investigate EVF generation inside liquid metals. Bojaverics et al. (3) analysed electrically induced flows in different geometries. He predicted that different structures of flow can evolve in the presence of an external magnetic field. Kharicha et al (4) performed experimental and numerical studies on the effect of MHD on flow and free surface deformation inside liquid metals confined in a spherical container. The study confirmed the extreme sensitivity of flow to the current magnitude and external axial magnetic field. Herreman et al. (5) analysed the effect of EVF inside liquid metal batteries considering single and multiphase flow. The study predicted considerable deformation and shortcuts at the interface at a

IOP Conf. Series: Materials Science and Engineering 1309 (2024) 012004 doi:10.1088/1757-899X/1309/1/012004

threshold current. Liu et al. (6) reached a good validation between experimental measurements and numerical simulation of EVF inside a cylindrical container. The effect of the earth's magnetic field in the simulation was necessary to match the experimental results. Kharicha et al. (1) conducted a study to cover a large range of applied current and axial external magnetic field inside a liquid metal. Different flow structures were predicted inside the cylindrical shape geometry. Considering no external fields, EVF structure was obtained. However, external magnetic transfer the flow structure to a tornado, cyclone, or inverted tornado depending on the strength of the magnetic field.

This study presents a 3D numerical model to predict MHD flows in presence of earth magnetic field. The study covers a wide range of applied currents from 50 A up to 1 kA. In the absence of external magnetic field, the effect of applied current is analysed on 3D flow structures. In the presence of an earth magnetic field, the influence of magnetic field on the flow is studied. The goal is to understand the EVF structures in 3D and assess the mixing effects of MHD flows in industrial processes.

2. Materials and methods

The system consists of a 3D cylinder containing electrically conducting liquid (steel). The cylinder is connected to two solid electrodes from the top (cathode) and bottom (anode). The bottom electrode covers the whole diameter of the cylinder while the top electrode is in contact with one-fifth of the total diameter. The top surface of the liquid is assumed to be flat with a slip condition. Figure 1 shows the configuration of the domain. Point A is the center of domain where the data for velocity is collected for analysis.

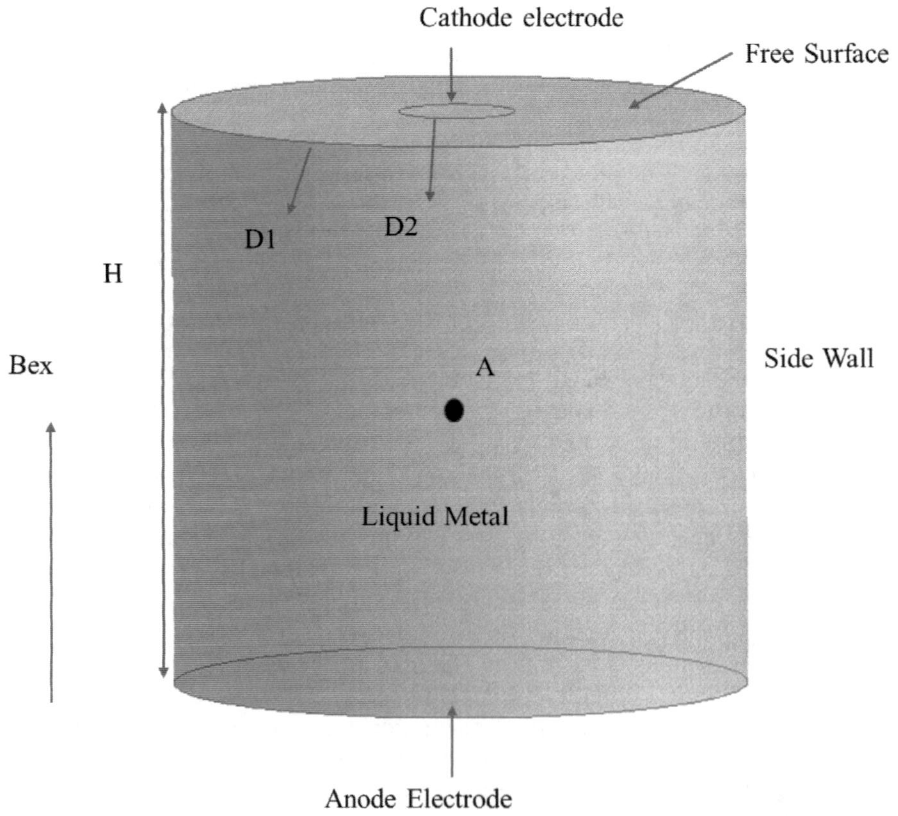

Figure 1. schematic representation of computational domain and boundaries. Dimensions are H = D1 = 30 cm D2 = 6 cm. Bex represents the direction of external magnetic field.

The height of cylinder (H) is equal to the diameter of cylinder (D1) with size of 30 cm, (D2) the smaller electrode diameter is equal to 6 cm. The current enters from the bottom electrode to the cylinder and exists from D2 the smaller electrode. The interface and side walls are insulated. A very fine mesh of 6.1 million elements is used to capture the detailed flow and the turbulence inside the domain. Isothermal flow inside the domain is considered. The continuity, flow (Navier-Stoke), and magnetic vector potential equations are solved in a coupled fashion (1). To account for the turbulence the LES model is used (7). The conducting liquid inside the cylindrical geometry is chosen to have liquid steel properties. The density of liquid steel is 7000 kg/m^3 and viscosity is 0.006 Pa.s.

3. Results

3.1. No external magnetic field:

The simulations were performed in a transient manner. Each case was simulated until a steady or quasi-steady flow is achieved. No external magnetic field was considered for the first part of the study. Figure 2 shows the velocity contours and streamlines of the final state. Considering low applied current (50 A and 100 A), a typical EVF flow develops and reaches a steady state. The Lorentz force generates the flow near the small electrode and a global jet pushes the flow downwards. For the case of 250 A, initially, an electro vortex is observed however the structure of the EVF is disturbed with frequent oscillations. Further increase in applied current leads to larger disturbance and more violent mixing (500 and 1000 A). At 1000 A the turbulence/disturbance can be observed in the whole domain. Some swirl flow is also observed however it was periodic and varies in direction.

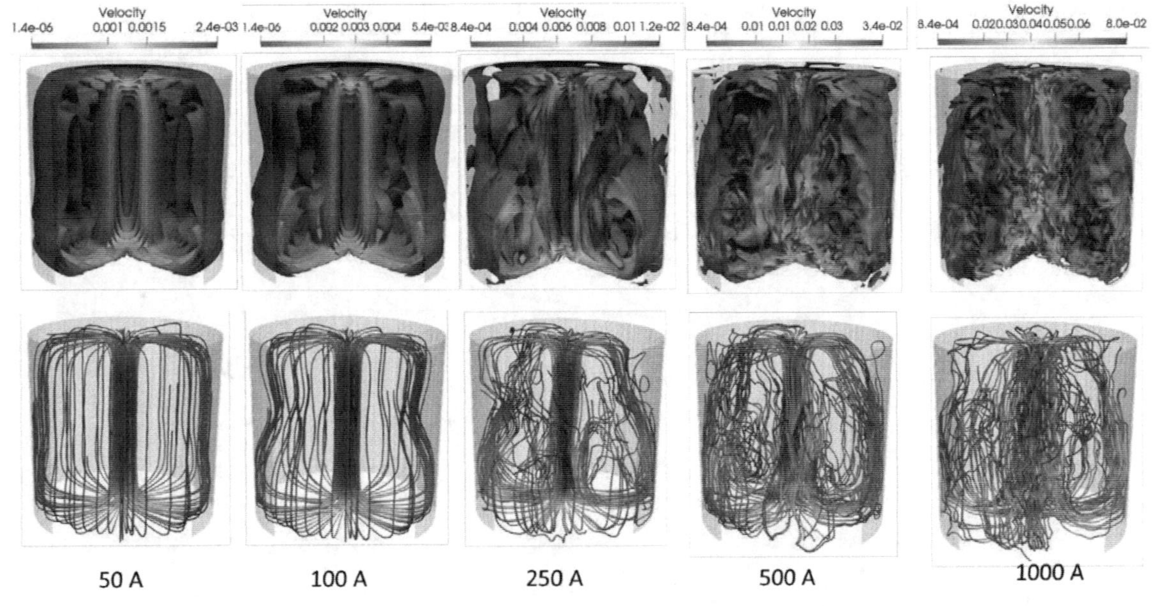

Figure 2. Velocity contour in the top and the flow streamlines inside the domain in the lower portion. Different current values are shown: 50, 100, 250, 500, and 1000 A.

The local kinetic energy (KE) at the center of the domain (point A in Figure 1) was recorded with time for different applied currents (Figure 3). For 50 A, the K.E reaches a constant value after a certain time and does not vary with time which indicates a steady state. At 250 A, after the EVF develops the KE varies with time slightly. The existing oscillations are small but visible. However, a further increase in the applied current enhances these discrepancies and induces strong turbulence (see KE for 500 A &

1000 A). As shown we also analyze the standard deviation (STD) for the KE at the same location (Figure 4). The data is collected every 0.25 sec. Considering low current, 50 A, the STD is very low. The STD increases by orders of magnitude when the applied current is higher. This indicates that these flows can be considered as chaotic and turbulent rather than unstable.

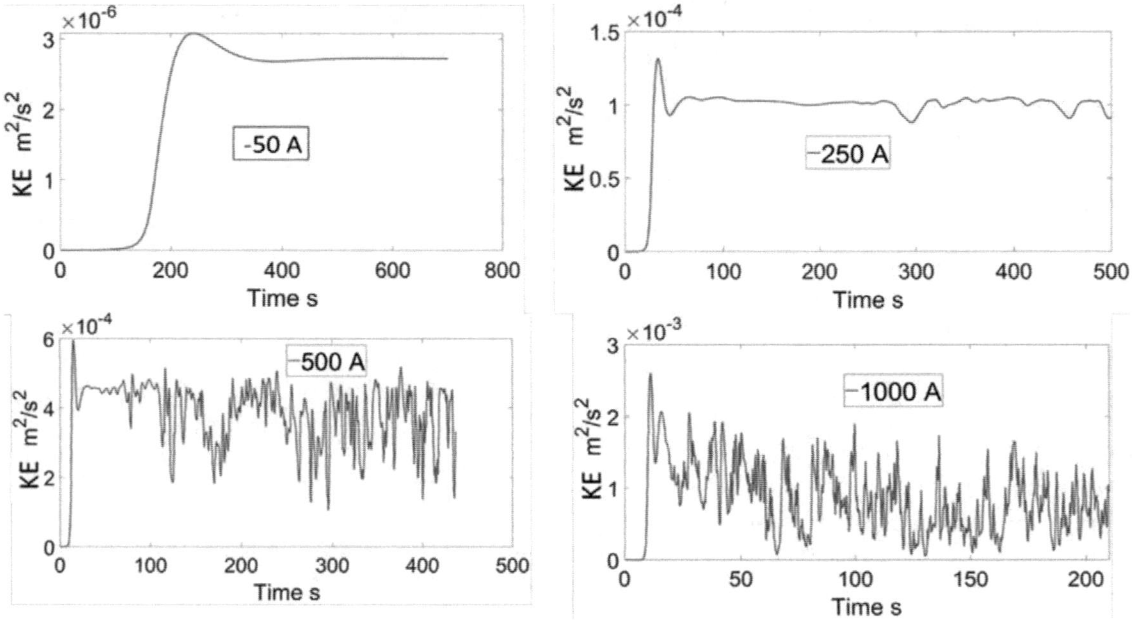

Figure 3. K.E. at the center of cylinder (A) as a function of time for different applied currents.

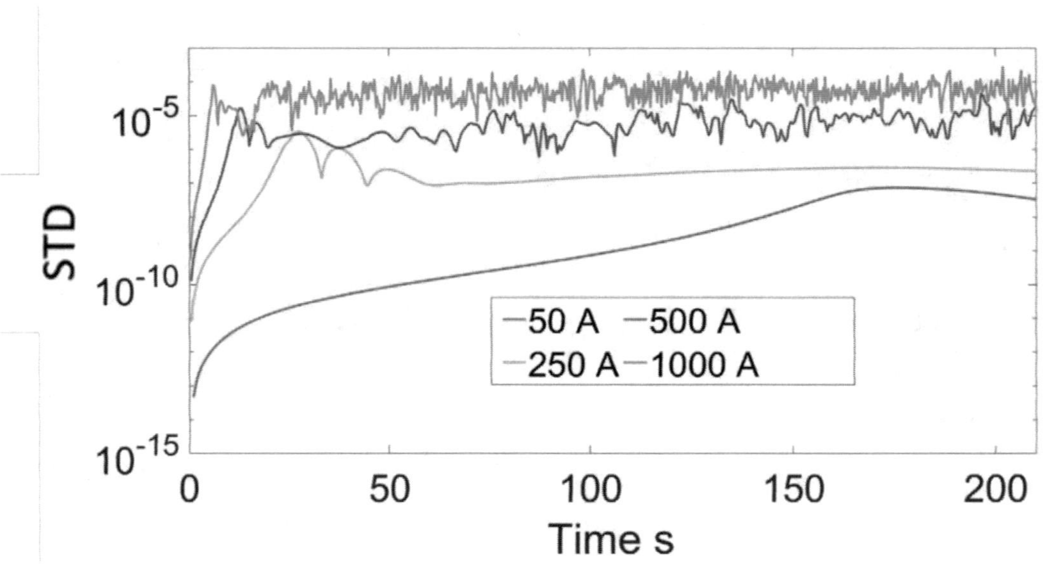

Figure 4. KE's standard deviation (STD) as a function of time at the center of the domain for different applied currents.

3.2. *Effect of external magnetic field:*

In the second part of the study, the axial external magnetic field was included into the model. The value of the external magnetic field is equal to the earth's magnetic field (50 μT) Figure 5. The flow is drastically affected by the axial magnetic field. For the lowest current 50 A, the flow is mostly in the azimuthal plane and a cyclone structure is predicted. The flow transfers into a tornado which narrows its diameter as current increase. The competing effect between EVF and azimuthal flow due to the external magnetic field can be clearly observed. Despite the transformation in structure due to the external magnetic field, the turbulence is still clearly visible when a high current is applied.

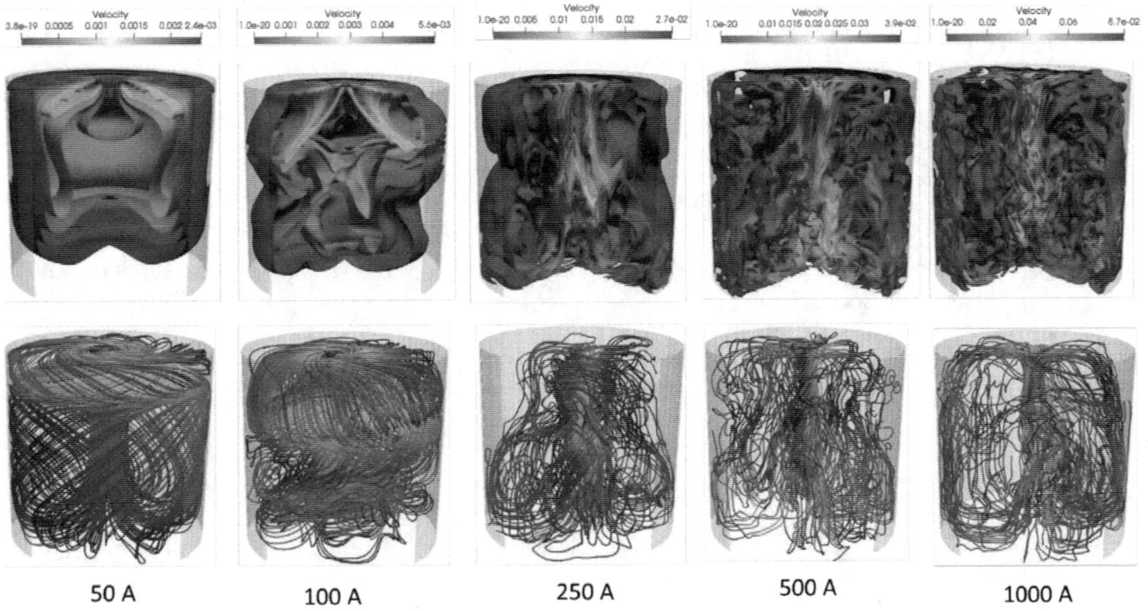

Figure 5. Velocity contours in the top part and the velocity streamlines inside the domain in the lower part. Different currents were used in the modeling: 50, 100, 250, 500, 1000 A. External magnetic field is considered B = 50 μT.

4. Conclusions

The study presents the structure of electrovortex flow (EVF) using 3D simulation in presence and absence of earth magnetic field. The simulations are performed inside liquid metal in a cylindrical geometry. In absence of earth magnetic field, low applied current leads to uniform electrovortex flow. However, we found that the increase in current leads to powerful oscillations and eventually turbulence as indicated by kinetic energy and standard deviation of the kinetic energy. The presence of earth magnetic field induces several structures like cyclone, tornado and rope tornado depending on the applied current.

5. Acknowledgments

The authors acknowledge the financial support by the Austrian Federal Ministry of Economy, Family and Youth and the National Foundation for Research, Technology and Development within the framework of the Christian Doppler Laboratory for Metallurgical Applications of Magnetohydrodynamics.

References

[1] Kharicha A, Al-Nasser M, Barati H, Karimi-Sibaki E, Vakhrushev A, Abdi M, et al. Tornados and cyclones driven by Magneto-hydrodynamic forces. Eur J Mech B/Fluids [Internet]. 2022;94:90–105. Available from: https://doi.org/10.1016/j.euromechflu.2022.02.001

[2] Shercliff JA. Fluid motions due to an electric current source. J Fluid Mech. 1970;40(2):241–50.

[3] Bojarevi V, Freibergs Y, Shilova EI, Shcherbinin E V. Electrically induced vortical flows. Vol.

9. Springer Science & Business Media; 1988.

[4] Kharicha A, Teplyakov I, Ivochkin Y, Wu M, Ludwig A, Guseva A. Experimental and numerical analysis of free surface deformation in an electrically driven flow. Exp Therm Fluid Sci. 2015;62:192–201.

[5] Herreman W, Nore C, Ramos PZ, Cappanera L, Guermond J-L, Weber N. Numerical simulation of electrovortex flows in cylindrical fluid layers and liquid metal batteries. Phys Rev Fluids. 2019;4(11):113702.

[6] Liu K, Stefani F, Weber N, Weier T, Li BW. Numerical and experimental investigation of electro-vortex flow in a cylindrical container. Magnetohydrodynamics. 2020;56(1):27–41.

[7] Nicoud F, Ducros F. Subgrid-scale stress modelling based on the square of the velocity. Flow Meas Instrum. 1999;62:183–200.

Coupled dynamic modeling of scrap melting and gas phase reactions in the electric arc furnace process

I Mäkelä[1], V-V Visuri[1], M Aula[2] and T Echterhof [3]

[1] Process Metallurgy Research Unit, University of Oulu, PO Box 4300, 90014 University of Oulu, Finland
[2] Outokumpu Stainless Oy, Terästie, 95400 Tornio, Finland
[3] Department for Industrial Furnaces and Heat Engineering, RWTH Aachen University, 52074 Aachen, Germany

ilpo.makela@oulu.fi

Abstract. Gas phase reactions have a significant effect on the composition and heat content of the off-gas in the electric arc furnace (EAF) process. In this work, a previously developed dynamic scrap melting and heat transfer model was coupled with a gas phase reaction module based on Gibbs energy minimization. The gas phase reaction module retrieves the necessary thermodynamic data from a previously developed thermochemistry module. The gas phase reaction module has been used to improve the description of the gas burners and the freeboard of the EAF. The implementation of the gas phase reaction module has been found to be in good agreement with commercially available software (HSC Sim) for calculations assuming a gas phase equilibrium.

1. Introduction

The state of the gas phase within the electric arc furnace (EAF) is an important factor when calculating the energy balance and efficiency of the process. Oxygen in the leakage air or lancing can react with unburnt components of the gas phase, but this leakage air can also cause significant cooling and heat loss in the process. Additionally, the outflow of gases, whether through the hood or gaps in the furnace, can remove significant amounts of unreacted components that could otherwise have contributed to heating the process. Monitoring the gas phase can also provide valuable insight into the progress of the melting and refining processes. Due to these factors, including the gas phase in EAF process modeling is an important issue to consider.

Hay et al. [1] have done an extensive review of fast and comprehensive EAF process models. In their review, it is found that most of the models developed in the last 10 years are based on the works of Logar et al. [2–7], or developed in cooperation with them. The description of the gas phase equilibria in these models mainly considers the combustion of CH_4 and CO, the water-gas shift reaction, the Boudouard reaction and the reaction of C_9H_{20} with C. These reactions are mainly modeled with the equilibrium constant method.

MacRosty and Swartz [8] modeled the gas phase as a zone where all the components are assumed to exist in a state of chemical equilibrium, which is calculated using the Gibbs free energy minimization method. The limiting factor of reactions is the mass transfer between the different zones, with the mass transfer coefficient being an adjustable parameter, which can be estimated from industrial data. Mass transfer in the gas phase considers the burners, leakage air, off-gas, electrode cooling with water and transfer occurring between the slag-metal interaction zone. The direction of the leakage air is related to the furnace pressure, but some amount of off-gas seems to always be removed.

The model developed by Meier et al. [9] is based on the model developed by Logar et al. [4,5], with an enhanced description of the off-gas. Meier pointed out that significant amounts of H_2O, H_2 and even CH_4 are measured in the off-gas, and therefore these were included in their model. In Logar's model these components were not included in the gas phase, and complete combustion of CH_4 was assumed. Meier uses equilibrium constants to calculate some of the reaction rates. These equilibrium constants were assumed to only be temperature dependent and were pre-calculated for a range of temperatures for each of the equilibrium reactions to improve model performance. Leak air calculation was done using relative furnace pressure and a constant relating the pressure to the mass flow rate.

The model developed by Hay et al. [10] is based on the model developed by Meier. In their work, Hay pointed out that their pressure calculation within the furnace is unstable, and can start to oscillate, which in turn can slow down or even crash the simulation. This oscillation is caused by a feedback loop from the coupling of the leak air ingress with the pressure and temperature calculations. To counteract this, Hay used empirically derived equations that allowed the model to detect steep changes in the leak airflow and adjust the time step to avoid oscillations.

Our aim is to improve the description of the gas phase. By applying the Gibbs energy minimization method to the description of the gas phase reactions, we could improve the overall accuracy and ease of implementation of even complex gas mixtures. The method does not require the user to define the occurring reactions and their stoichiometries, so implementing new burner gases and accounting for side reactions is quite simple. Additionally, the method inherently considers the interactions between the reactions, while the equilibrium constant method minimizes the Gibbs energy only for individual reactions. In the present paper, we describe the coupling of the previously developed gas phase reaction module [11] and dynamic EAF scrap melting and heat transfer module [12]. The long-term goal is to also consider and implement kinetics into the gas phase modeling for more accurate results.

2. Materials and methods

The scrap melting and heat transfer module originally developed by Ringel [12] is based on previous EAF model work by Logar et al. [3–5], Meier [13] and Fathi [14], and focuses on describing the scrap melting and heat transfer processes. The model uses the explicit Runge-Kutta-Fehlberg method to find the solution using adaptive time-steps. The model reads process data to determine the electrical energy and burner energy inputs, which are then distributed according to a calculated heat transfer matrix. This energy distribution is then used to calculate the change of temperature in different zones as well as the change in mass of solid scrap, melt, slag, etc. In the original implementation, the conversion of the gas burner was assumed to be 100%, and the use of the gas burner did not change the composition of the gas phase, which was simplified to be at the composition of normal atmospheric air. The heat from the burner is transferred to the solid and liquid scrap and the gas phase according to a hyperbolic-tangent approximation also used by Logar et al. [4]:

$$Q_{\text{CH4-sSc}} = Q_{\text{r-CH4}} K_{\text{burn}} \left[0.35+0.65\tanh\left(\frac{1300°\text{C}}{T_{\text{sSc}}-273.15\text{K}}-1\right)\right] \tag{1}$$

$$Q_{\text{CH4-gas}} = Q_{\text{r-CH4}} \left\{1-K_{\text{burn}} \left[0.35+0.65\tanh\left(\frac{1300°\text{C}}{T_{\text{sSc}}-273.15\text{K}}-1\right)\right]\right\} \tag{2}$$

where $Q_{\text{CH4-sSc}}$ is the heat transfer to the scrap [W], $Q_{\text{CH4-gas}}$ is the heat transferred to the gas phase [W], $Q_{\text{r-CH4}}$ is the heat released in the burning [W], K_{burn} is burner efficiency, which is approximated to be 0.7 based on Ref. [15], and T_{sSc} is the temperature of the solid scrap [K].

The gas phase reaction module developed by Jussila [11] uses the minimization of Gibbs energy method, described by White et al. [16] The minimization method retrieves the Gibbs energy data for the species involved from our thermochemistry module. In addition to participating gas species, initial amounts, temperature, and pressure, the equilibrium module can be given a parameter to limit the portion of the input gases, that partake in the equilibrium calculation. By adjusting this parameter, we can force the model to decrease the maximum conversion achieved. In practice this is done by multiplying the molar input values by the limit parameter, which ranges between 0 to 1 for each molecule, and this portion is then fed to the equilibrium calculation routine. The leftover moles of the previous multiplication pass the equilibrium calculation, effectively making them passive from the reactions' standpoint. The leftover moles are then summed with the equilibrated moles to get the final molar amounts.

The thermochemistry module developed originally in C++ in Ref. [17] was converted to Python in Ref. [12] to provide thermochemical properties of different species and their mixtures, as well as reactions at a range of temperatures. The module retrieves tabulated data such as the Shomate coefficients, standard enthalpy and standard entropy from a csv file, and processes this into a Python dictionary, from which the equations (functions) can then retrieve the needed data with species name, temperature, and property index as parameters. If thermochemical data is not available for a given temperature, the model extrapolates using the last available values, though for the present work, we have enough data to cover the operating temperatures of the EAF. For calculating the Gibbs energies, the thermochemistry module uses the following equations:

$$C_{\text{p}} = A + B\cdot T\cdot 10^{-3} + C\cdot T^{-2}\cdot 10^{5} + D\cdot T^{2}\cdot 10^{-6} + E\cdot T^{-3}\cdot 10^{8} + F\cdot T^{3}\cdot 10^{-9} \quad (3)$$

$$\Delta H = H^{\circ}_{298.15} + \int_{298.15}^{T} C_{\text{p}}\,\mathrm{d}T + \sum H_{\text{tr}} \quad (4)$$

$$\Delta S = S^{\circ}_{298.15} + \int_{298.15}^{T} \frac{C_{\text{p}}}{T}\,\mathrm{d}T + \sum S_{\text{tr}} \quad (5)$$

$$\Delta G = \Delta H - T\Delta S \quad (6)$$

where C_{p} is the molar heat capacity [J K^{-1} mol^{-1}], A through F are coefficients for the Shomate equation, ΔH is the molar enthalpy at temperature T [J mol^{-1}], $H^{\circ}_{298.15}$ is the standard enthalpy [J mol^{-1}], $\sum H_{\text{tr}}$ is the enthalpy associated with phase transformations occurring between temperatures 298.15 K and T [J mol^{-1}], ΔS is the molar entropy at temperature T [J K^{-1} mol^{-1}], $S^{\circ}_{298.15}$ is the standard entropy [J K^{-1} mol^{-1}], $\sum S_{\text{tr}}$ is the entropy associated with phase transformations occurring between temperatures 298.15 K and T [J K^{-1} mol^{-1}] and ΔG is Gibbs energy at temperature T [J mol^{-1}].

The implementation of the gas burner from Logar et al. [4,5] was modified by using the gas phase equilibrium module to calculate the equilibrium composition of the gas leaving the burner, and the energy released in the reactions. This allows us to model the conversion of the burner and possible side reactions with a relatively simple implementation. The calculation routine takes the possible gas species present in the burner region, the fuel gas feed rate, and gas phase temperature as inputs. It then calculates the stoichiometric amount of oxygen needed for complete burning of the fuel gas and adds it to the reaction. The module then calculates the equilibrium at the given gas phase temperature, and the change in enthalpy to find the energy released, which is then divided between the scrap and the gas phase according to equations (1) and (2). The calculation also gives the feed rate for output gases, which are added to the gas phase according to the taken time step size in the solver. The implementation allows for a quick and easy way to change the used fuel and lambda (supplied amount of oxygen compared to the stoichiometric amount of oxygen) in the simulation.

The equilibrium module was also implemented to consider the whole freeboard gas phase of the EAF. In its current implementation, the gas phase is treated as an adiabatic equilibrium reactor. The model uses the secant method to find a temperature and composition where the error between the initial and final enthalpy is below the set tolerance. The leakage air is adjusted by calculating the maximum amount of gas that can be present in the freeboard, using the ideal gas law, and comparing this with the current amount of gas in the calculation. If there is more gas than the freeboard can contain in its current conditions, the excess leaks out, but if there is less gas, air leaks in diluting the composition and lowering its temperature, further increasing the amount of gas leaking in. The balance between the new temperature and air intake is solved using the secant method. The calculation routine is illustrated in figure 1. The present implementation enables initial testing on the gas phase module. In further work, it would be relevant to modify the model to account for a pressure-driven gas flow to obtain more realistic results.

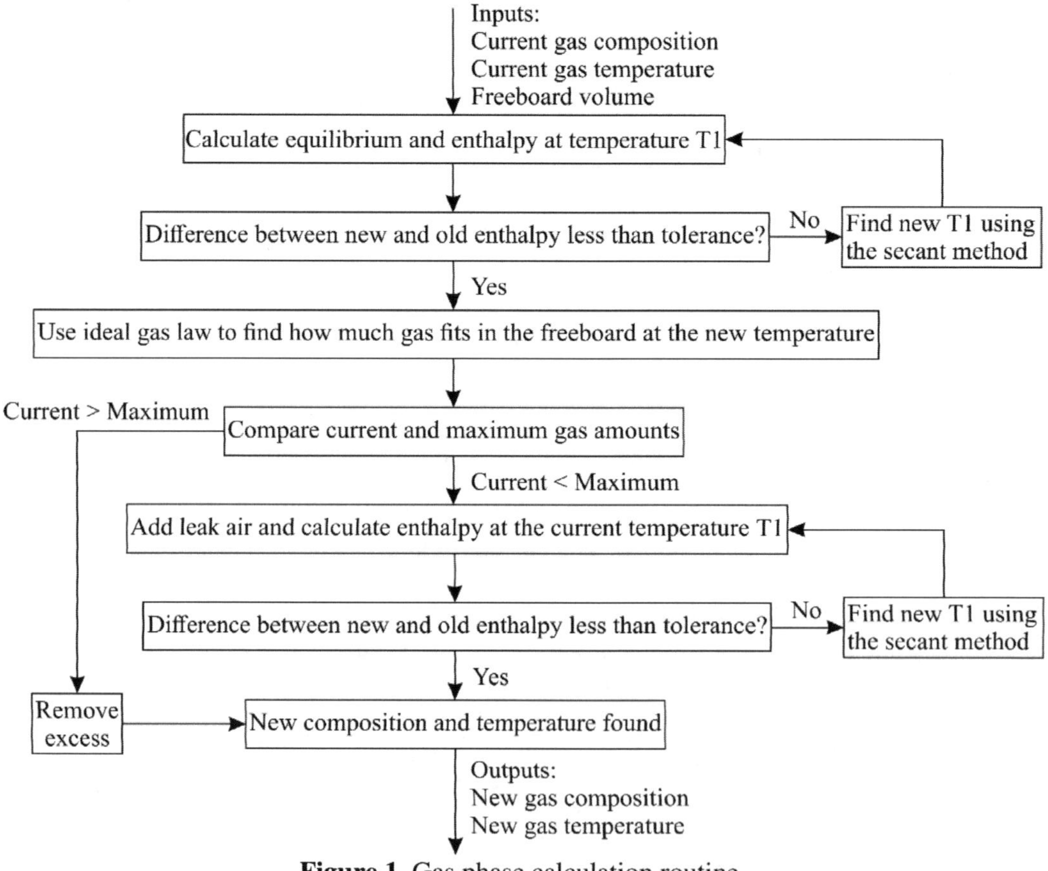

Figure 1. Gas phase calculation routine.

3. Results

Our implementation of the gas phase equilibrium calculation was compared with the HSC Chemistry 10 (version 10.0.7.13) Sim module equilibrium calculation, through the burning of methane. If the amount of some gas specie(s) reaches zero in the simulation, it can lead to the model trying to calculate the natural logarithm of zero at these points. To counteract this, the model checks for these zero values, and replaces them with a very small value of 10^{-10} moles. All the following simulations were performed on an laptop with 11th Gen Intel(R) Core(TM) i5-1135G7 @ 2.40GHz processor and Windows 10 64-bit operating system.

The simulated burner was fed with 10 moles of CH_4 and 500 moles of air (79 mol-% N_2, 21 mol-% O_2) at 1500 °C (1773.15 K). The output species specified for the system were N_2, O_2, CH_4, CO, CO_2, $H_2O_{(g)}$ and H_2, while the temperature of the output was controlled so that an energy balance was achieved between the inputs and outputs. The results of the simulations are presented in table 1. The results of the simulations using the HSC Sim and Python models are in nearly perfect agreement. The gas phase compositions are almost identical, and the temperature difference is only 0.40 °C, while the increase in temperature was around 422 °C.

For the studied system, the calculation time to find the equilibrium composition and temperature was approximately 0.8 s using an error tolerance of 10^{-5} J. The input and output states are far apart, meaning longer calculation times to find the equilibrium. When the input and output states are close together, for example in a simulation where the modeled gas phase changes little between taken time steps, the calculation times can be assumed to be much faster. HSC is generally able to find the equilibriums much faster than our Python model, however, HSC cannot be coupled with the Python model or used in online process modeling.

Table 1. Equilibrium calculation for HSC sim model and Python model.

			Output	
Property	Unit	Input	HSC sim model	Python model
CH_4	mol	10	0	$6.70 \cdot 10^{-20}$
N_2	mol	395	395.00	395.00
O_2	mol	105	85.11	85.10
CO	mol	0	0.14	0.14
CO_2	mol	0	9.86	9.86
$H_2O_{(g)}$	mol	0	19.94	19.95
H_2	mol	0	0.054	0.054
Temperature	°C	1 500	1922.60	1922.20

Additionally, the sensitivity of the EAF model coupled with the equilibrium module was inspected by adjusting model parameters such as maximum time step size (Δt), and relative and absolute tolerances of the solver. It should be noted that as the model uses adaptive Δt, the modeled points in time are rarely exactly the same, and the compared points need to be chosen so that they are close to each other. Inspecting the table 2, containing the results from adjusting the maximum Δt, decreasing the Δt value below 1 s, seems to change the results very little. Increasing the Δt to 2s changes the temperature by 7.05 K and 8.16 K compared to 1 s Δt, at simulation points 920 s and 2 000 s respectively. In tables 3 and 4, we have compiled the results from adjusting the relative and absolute tolerances of the solver. Decreasing the tolerance values below 1 changes the results of the model very little. It was chosen that 1 s Δt, and tolerance values of 1 are adequate for the use of the model.

IOP Conf. Series: Materials Science and Engineering 1309 (2024) 012005 doi:10.1088/1757-899X/1309/1/012005

Table 2. Model gas phase sensitivity to the Δt size, with tolerance values set to 1.

max Δt	Model time (s)	Gas temperature (K)	Mass of gas (kg)	CO_2 amount (mol)
2.0 s	920.67	1 378.06	12.89	161.06
1.0 s	919.99	1 385.11	12.82	159.47
0.5 s	920.01	1 384.47	12.84	159.69
0.2 s	920.09	1 384.24	12.84	159.76
2.0 s	2 000.71	1 475.78	10.91	136.20
1.0 s	2 000.14	1 483.94	10.86	135.19
0.5 s	2 000.24	1 480.37	10.88	135.75
0.2 s	2 000.16	1 480.75	10.88	135.66

Table 3. Model gas phase sensitivity to the relative tolerance of the solver, with Δt = 1 s and absolute tolerance set to 1.

Relative tolerance	Model time (s)	Gas temperature (K)	Mass of gas (kg)	CO_2 amount (mol)
1.0	919.99	1 385.11	12.82	159.47
0.5	919.99	1 385.17	12.82	159.47
0.2	920.10	1 383.79	12.84	159.70
1.0	2 000.14	1 483.94	10.86	135.19
0.5	2 000.17	1 483.74	10.86	135.13
0.2	2 000.15	1 483.56	10.85	135.09

Table 4. Model gas phase sensitivity to the absolute tolerance of the solver, with Δt =1 s and relative tolerance set to 1.

Absolute tolerance	Model time (s)	Gas temperature (K)	Mass of gas (kg)	CO_2 amount (mol)
1.0	919.99	1 385.11	12.82	159.47
0.5	919.79	1 385.05	12.82	159.69
0.2	919.97	1 385.13	12.82	159.47
1.0	2 000.14	1 483.94	10.86	135.19
0.5	2 000.18	1 483.81	10.86	135.13
0.2	1 999.90	1 482.04	10.87	135.58

Additionally, the effect of adjusting the conversion of the burner was studied. This was conducted by decreasing the amount of oxygen supplied with the fuel gas, and by adjusting the limiter parameter. In the first case, the amount of oxygen supplied was reduced to 90% of the amount required for the complete burning of the methane gas. The results are shown in figure 2. The supplied methane first exhausts the oxygen in the gas phase, after which CO and H_2 start to form due to the lack of O_2. The lack of oxygen leads to incomplete combustion of CH_4 according to equation (7):

$$CH_4 + \frac{1}{2}O_2 \rightarrow CO + 2H_2 \tag{7}$$

In this case, however, we would be seeing H_2 to CO rates of 2:1, which is not the case in our simulation. At the beginning, we observe higher than expected amounts of H_2, and at a later stage we observe higher than expected amounts of CO. This can be explained by the water-gas shift reaction shown in equation (8):

$$CO + H_2O \rightarrow CO_2 + H_2 \tag{8}$$

It is likely that at the beginning the thermodynamic conditions favor the formation of reaction products (CO_2 and H_2), but as the gas phase temperature rises, the reaction equilibrium shifts to favor the formation of reactants (CO and H_2O). However, we need to state that we don't know the exact reactions occurring in our system, as defining them is not needed for calculating the equilibrium compositions using the Gibbs energy minimization method.

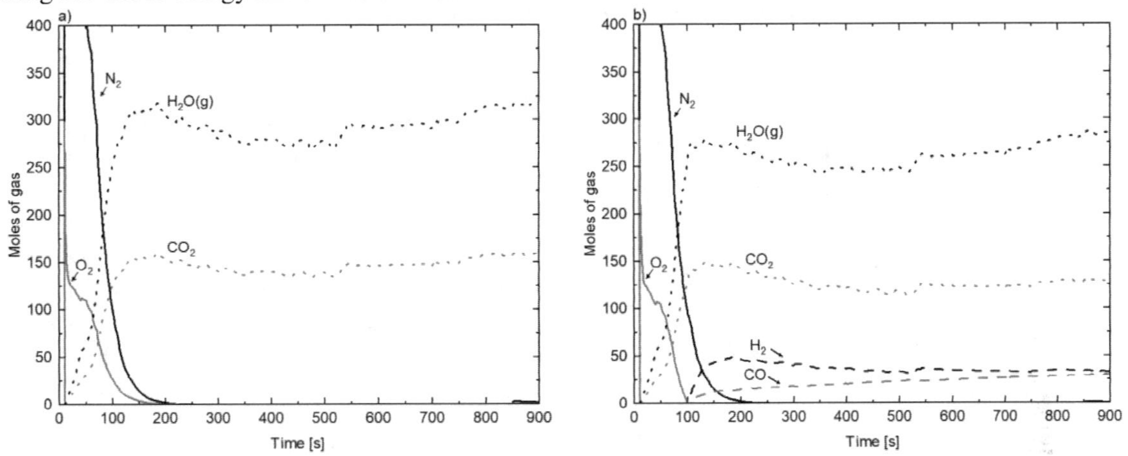

Figure 2. Composition of the gas phase when the burner is supplied with a) stoichiometric amount of oxygen, b) 90% of the stoichiometric amount of oxygen.

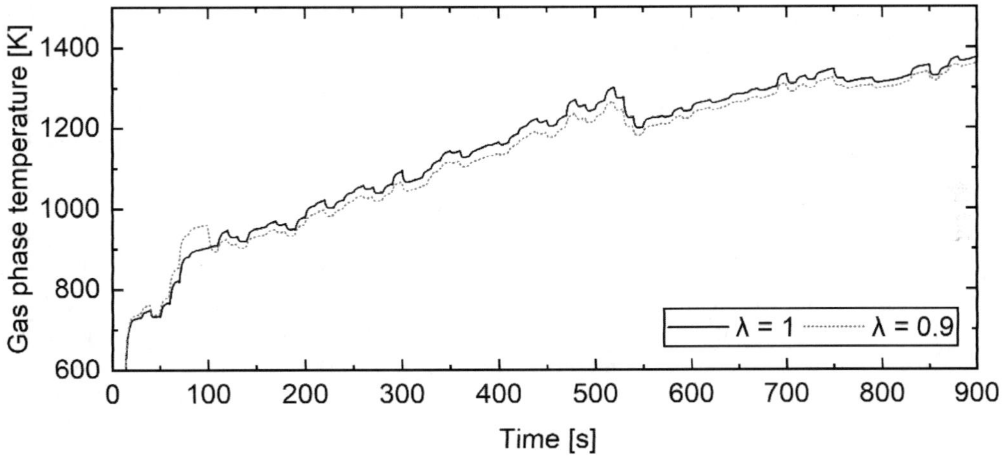

Figure 3. Gas phase temperature when the burner is supplied with stoichiometric amount of oxygen ($\lambda = 1$) and 90% of the stoichiometric amount of oxygen ($\lambda = 0.9$)

Also, the temperature of the gas phase was on average 16 K hotter in the case with stoichiometric amount of oxygen, as shown in figure 3. In the figure we can also see that at the beginning the gas phase is significantly hotter in the case with limited oxygen supply. This is caused by the way that heat is distributed in the burner calculation and will be expanded upon in the following paragraph.

When the burner was supplied with a stoichiometric amount of oxygen, but the reaction was limited to 75% conversion, the gas phase was found to be generally hotter than in the case with 100% burner conversion (see figure 4). This can be attributed to the way that the energy from the burning is divided. When the methane is burned in the burner, the energy is divided between the gas phase and the metal. However, when the combustion is not complete, but we supply the furnace with the required oxygen, the methane reacts in the freeboard gas phase equilibrium calculation, where the energy is not divided, instead staying in the gas phase and then transferred to the other parts of the furnace according to the heat transfer matrix. This is also what happens at the beginning of our previous example (λ = 0.9 in figure 3), where some of the burner gases react with the oxygen in the freeboard.

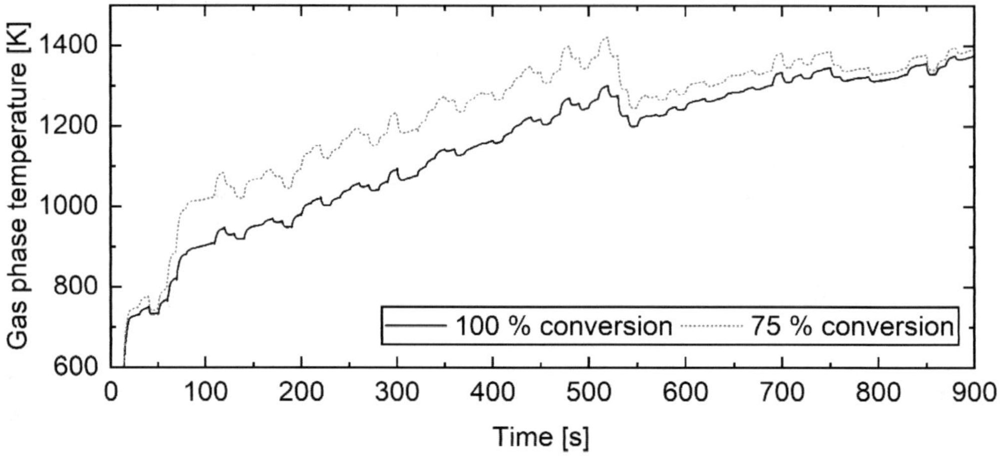

Figure 4. Gas phase temperature comparison

4. Discussion

The implementation of the gas phase reaction module has allowed the previously developed scrap melting and heat transfer module to also consider the composition and reactions of the gas phase, with a relatively simple implementation. The implementation however requires further improvements for more realistic modeling results, especially for the description of the leakage air. Currently, the leakage air is modeled in a way, that rarely allows air to leak in, hindering some reactions and the heat losses caused by this leakage air. The implementation also neglects to consider the kinetics of gas phase reactions, instead assuming the gas phase to always be in a state of equilibrium.

Using the gas phase reaction module at each taken time-step also increases the computation time to a degree where the model is not usable in online process modeling applications. To improve the computation times, either the gas phase reaction module needs to be optimized to find the equilibrium faster, or its use frequency needs to be lowered. The implementation could be further improved by having the gas phase reaction module find the equilibrium after pre-determined time steps or changes in the system and have reaction kinetics then drive the composition towards the found equilibrium.

5. Conclusions

The present paper describes the first implementation of coupling previously developed models for scrap melting and gas phase reactions. The implementation allows a more detailed description of the gas phase composition and the reactions occurring in it. The results obtained using the gas phase reaction module were found to be in excellent agreement with commercially available software. Sensitivity analysis indicates that the coupled model is sufficiently robust for process modeling. As for further work, the recommended improvements include the implementation of reaction kinetics, improvements to the implementation of the leakage air, and optimization of the equilibrium calculation and its implementation.

6. Acknowledgements

This work was conducted within the framework of the Towards Fossil-Free Steel (FFS) project funded by Business Finland and the Towards CO_2-lean electric arc furnace steelmaking through fundamental and data-driven mathematical modeling (LEAF) project funded by the Research Council of Finland (project no. 356439). Furthermore, Technology Industries of Finland Centennial Foundation Steel and Metal Producers' Fund is gratefully acknowledged for the financial support of Ilpo Mäkelä.

References

[1] Hay T, Visuri V-V, Aula M and Echterhof T 2021 A Review of Mathematical Process Models for the Electric Arc Furnace Process *Steel Res Int* **92** 2000395
[2] Logar V, Fathi A and Škrjanc I 2016 A Computational Model for Heat Transfer Coefficient Estimation in Electric Arc Furnace *Steel Res Int* **87** 330–8
[3] Logar V and Škrjanc I 2012 Modeling and Validation of the Radiative Heat Transfer in an Electric Arc Furnace *ISIJ Int* **52** 1225–32
[4] Logar V, Dovžan D and Škrjanc I 2012 Modeling and Validation of an Electric Arc Furnace: Part 1, Heat and Mass Transfer *ISIJ Int* **52** 402–12
[5] Logar V, Dovžan D and Škrjanc I 2012 Modeling and Validation of an Electric Arc Furnace: Part 2, Thermo-chemistry *ISIJ Int* **52** 413–23
[6] Logar V, Dovžan D and Škrjanc I 2011 Mathematical Modeling and Experimental Validation of an Electric Arc Furnace *ISIJ Int* **51** 382–91
[7] Logar V and Škrjanc I 2012 Development of an Electric Arc Furnace Simulator Considering Thermal, Chemical and Electrical Aspects *ISIJ Int* **52** 1924–6
[8] MacRosty R D M and Swartz C L E 2005 Dynamic Modeling of an Industrial Electric Arc Furnace *Ind Eng Chem Res* **44** 8067–83
[9] Meier T, Gandt K, Echterhof T and Pfeifer H 2017 Modeling and Simulation of the Off-gas in an Electric Arc Furnace *Metall Mater Trans B* **48** 3329–44
[10] Hay T, Echterhof T and Visuri V-V 2019 Development of an Electric Arc Furnace Simulator Based on a Comprehensive Dynamic Process Model *Processes* **7** 852
[11] Jussila R *Modelling of gas-phase reactions in the freeboard of the EAF* Unpublished master's thesis (Oulu: University of Oulu)
[12] Ringel A 2020 *Online modelling of an electric arc furnace* Master's thesis (Aachen: RWTH Aachen university)
[13] Meier T 2016 *Modellierung und Simulation des Elektrolichtbogenofens* PhD Thesis (Aachen: RWTH Aachen university)
[14] Fathi A, Saboohi Y, Škrjanc I and Logar V 2017 Comprehensive Electric Arc Furnace Model for Simulation Purposes and Model-Based Control *Steel Res Int* **88** 1600083
[15] Toulouevski Y N and Zinurov I Y 2013 *Innovation in Electric Arc Furnaces* (Springer)
[16] White W B, Johnson S M and Dantzig G B 1958 Chemical Equilibrium in Complex Mixtures *J Chem Phys* **28** 751–5
[17] Visuri V-V, Sulasalmi P, Vuolio T, Paananen T, Haas T, Pfeifer H and Fabritius T 2019 Mathematical Modelling of the Effect of Reagent Particle Size Distribution on the Efficiency of Hot Metal Desulphurisation *Proceedings of the 4th European Steel Technology and Application Days* (Düsseldorf, Germany)

IOP Conf. Series: Materials Science and Engineering 1309 (2024) 012006 doi:10.1088/1757-899X/1309/1/012006

New methodology for estimating the influence of Cu content of Ferrous Scrap materials in the Total Cost of Operation (TCO) of scrap mix in Electric Arc Furnace steelmaking

E. Nuñez[1], H. Priyadarshi[2], M. Picco[3], A. Vicente[1]
[1]ArcelorMittal Global Research and Development Spain, Chavarri nº 6 48910 Sestao (Bizkaia), Spain
[2]ArcelorMittal Global Research and Development - Maizières Process. Voie Romaine, BP 30320, 57283 Maizières-lès-Metz Cedex, France
[3]ArcelorMittal Europe Long Products.66, rue de Luxembourg L-4221 Esch-sur-Alzette, Luxembourg

Corresponding Author: asier.vicente@arcelormittal.com

Abstract. Scrap is the main raw material of the electric arc furnace (EAF) route of steelmaking and an important one in the primary route. The decarbonisation commitment acquired by the most relevant steelmaking companies, will only increase the amount of scrap used by both routes, making it a critical component to succeed in the worldwide challenge of reducing CO_2 emissions. While being a very relevant material, scrap characterisation is complex due to its heterogeneity, but furthermore, its quality is worsening as its demand increases. This includes the sterile content, such as copper. Higher sterile content scrap comes at a lower price, while scrap with a lower content in undesired elements is most costly. Being able to balance the cost and benefit of purchasing scrap with a higher or lower content of copper with a respective, lower or higher purchase cost would be of interest to optimise the mix and produce the required quality steel at the lowest possible cost. Nowadays, this cost benefit analysis can only be done by performing Total Cost of Operation (TCO) calculations that require a considerable amount of information, including but not limited to the complete chemical composition of scrap, market parameters, production forecast and other relevant data; moreover, involving an arduous optimization process. Based on a parametric study of the TCO calculation, this work proposes a mathematical model to estimate the impact on the TCO of a predefined scrap mix when only modifying the specifics of one given scrap type. The results would facilitate the decision making by comparing the two scenarios as a factor of cost. This mathematical model can be easily programmed and requires only 5 parameters to be run, while providing a low error result.

1. Introduction

Throughout the history of humankind, steel can be considered as one of the prime global industrial materials; it was initially produced from an abundant raw material (iron ore) which can be mass produced at a relatively low cost and easily transformed into specific products for further assemblage or final use.

Looking at 2021 figures [1], more than 1.9 billion tons of crude steel were produced worldwide by using one of the two steelmaking routes presented in Figure 1, producing around 7 – 9% of the world's direct industrial CO_2 emissions. Within this scenario, the electric arc furnace (EAF) route contributed to more than 560 million tons. This is of great importance considering sustainability and the challenges imposed by the aggressive global economic condition since 2008. The steelmaking industry is putting extensive effort in the continuous development of new steel products with higher performance and lower environmental footprint. This situation has placed the scrap recycling steel manufacturing processes as one of the main alternatives for the steelmaking industry.

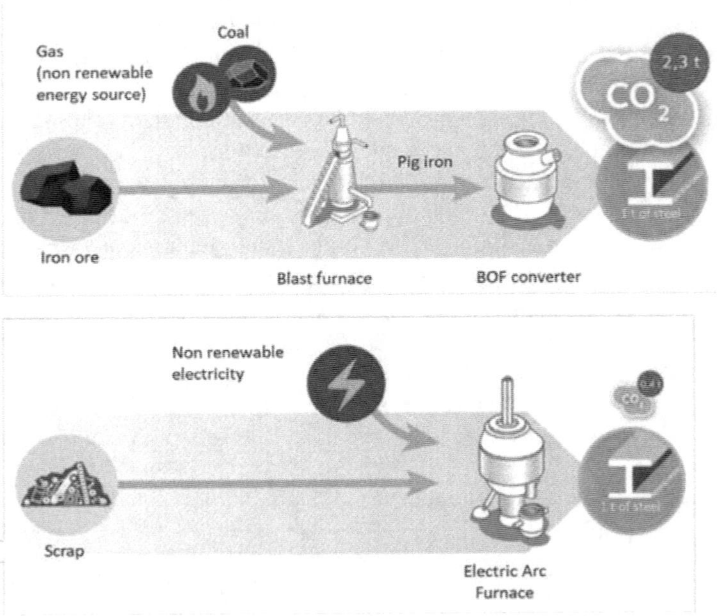

Figure 1, Global Picture of Steelmaking Process [2]

However, the management of the integrated steelmaking industries are keenly aware of current global environmental issues and are implementing practices that will reduce the hot metal ratio (CO_2 footprint). This would aim to increase the recycling of high-quality scrap metal that is superior to pig iron in many cases.

Industrial trends, likewise, show an increase of obsolete scrap availability (both in quantity and in complexity), coupled with greater scrap processing capacity in the short term, as shown in Figure 2.

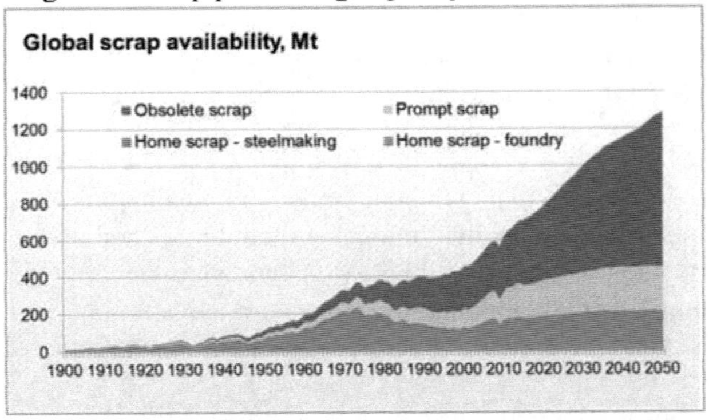

Figure 2, Estimation in Global Scrap availability evolution [3]

This new context places additional pressure on the electric steelmaking routes for two main reasons. Firstly, the availability of high-quality scrap will be lower in the future for EAF steelmaking, so in regions such as Europe, where the market for alternative ferrous materials, such as Hot Briquetted Iron (HBI) and Direct Reduced Iron (DRI) is limited, high quality scrap availability will be low. Therefore, implementing scrap mixture definition strategies to ensure the efficient use of ferrous material will be crucial for ensuring the sustainability of steelmaking processes within this new scenario. Secondly, the constituents present in low quality scrap materials are complex. Both the physical and the chemical characteristics of steel scrap that can come from a wide array of sources (ship and car breaking, rail tracks, rebars, turnings, etc.) are immensely heterogeneous and it may have undergone several processing steps before delivery to the scrap recycling plant or steel mill.

In an ideal situation, the material introduced into the furnace should only be steel. However, scrap materials are a mixture of steel with other types of materials that, in some cases have an adverse effect on the steel smelting process (loss of metallic performance, increased slag volumes, to name a few) and in other cases benefit the smelting process (such as reduction of energy consumption and alloy materials). However, optimal process management implies that the chemical composition of all the materials introduced into the furnace must be precisely controlled.

The concept of Value in Use (VIU) [4] has proven to be very useful when relating scrap quality with the operating cost of EAF operations. This concept includes a criterion for a comparative interpretation of the results generated by each scrap quality grade in the steelmaking process.

Apart from the purchasing costs, the VIU of a particular scrap grade will incur additional costs associated with the extra energy consumption and other materials such as electrodes, refractories and fluxes, due to the melting of non-ferrous materials included in regular scrap. The influence of different elements on EAF performance is depicted in Figure 3 where the impact of replacing 1% of Fe by 1% of referenced element is represented in colour code (Green colour represents positive impact on EAF performance and red colour represent negative impact). For the case of metallic yield, the impact of substituting Fe for any other compound is always negative and the colour code is ranging the negative impact (dark green means low negative impact and dark red means high negative impact).

		Metallic yield	Energy (kWh/t scrap)	Carbon (kg/t scrap)	Burnt Lime (kg/t scrap)	Slag, Refr. (kg/t scrap)
% Fe + tramp elements	Scrap with 100% Fe					
% C	+1% C -1%Fe					
% Al	+1% Al -1%Fe					
% Si	+1% Si -1%Fe					
% H	+1% H -1%Fe					
% acid gangue	+1% acid qangue -1%Fe					
% basic gangue	+1% basic gangue -1%Fe					
% O on Fe	+1% O -1%Fe					
% H2O	+1% H2O -1%Fe					

Figure 3, Influence of different elements on the EAF process performance [5]

Amongst all these non-desired compounds, the sterile fraction probably represents the most complex part in scrap and, the existing methodologies for on-site characterization of this fraction are still barely sufficient for full optimization of furnace performance.

Nevertheless, the conventional VIU approach only considers the thermodynamic impact of scrap compounds, meaning that the impact of all the constituents present in scrap that have a metallurgical influence on the EAF energy and mass balances. The practical limitation of the VIU concept is clearly demonstrated for the case of the Cu content in scrap materials. Two scraps with the same chemical composition, except for the Cu content, with the same purchasing price will offer the same VIU. However, the market value of the scrap with the lower Cu content is higher than the market value of the scrap with higher Cu content. It proves that the VIU criteria is not enough for establishing an overall material–by–material comparison methodology. This paper presents a complementary approach to assess the economic impact of Cu present in ferrous scrap materials on the scrap mix TCO.

2. The importance of scrap purchasing decisions in steelmaking

Around 85% of the total costs incurred by the EAF steel plants are associated with the activities of the steel shop [6]. Focusing on the steel shop, ferrous scrap accounts for about 50 to 60% of the total production cost in the EAF route (it will have an ever-increasing importance in the BOF route as progress is made towards the decarbonisation of the process) and 40% of the cost is related to operational parameters, mainly the metallic yield.

However, the real value of scrap does not only represent its iron content and price. Thus, when the potential feasibility of using a particular scrap grade is analysed, other factors apart from the purchasing price need to be looked at.

For normal electric steelmaking operation, a mixture of different scrap grades is periodically defined as a regular scrap charging mix. This mix is selected to optimally reduce the total manufacturing cost while

IOP Conf. Series: Materials Science and Engineering 1309 (2024) 012006 doi:10.1088/1757-899X/1309/1/012006

accomplishing the physical and technical facility requirements. For that, the Total Cost of Ownership (TCO) concept is used.
According to ArcelorMittal [7], TCO is be defined as: "A holistic mindset and analytical approach to purchase goods and services that incorporates not just price, but all costs associated to acquisition, transport, storage, usage and disposal". It allows the investigation and reduction of all associated costs of purchased goods by simultaneously pulling all levers: commercial (supplier consolidation, volume bundling, negotiations), demand (standardization, substitution, mix optimization) and process (inventory optimization and transport optimization).
Based on the TCO approach, the importance of the scrap quality concept lays on the necessity to optimize process variable costs by understanding the interrelations among the different aspects of scrap materials (physical specifics of the scrap materials, chemical composition of scrap material and conjunctural market situation) over the operating results of the melting process. These concepts are represented in Figure 4 in the scrap mix definition methodology side.
This TCO is always constrained by some physical and chemical conditions that must be accomplished by the whole of the scrap mixture, which are defined by the specifics of the product and the process. The deep understanding of the concept described above would allow determining the scrap mix and the scrap bucket layering to be sent to the steelshop as shown on the right side of Figure 4.

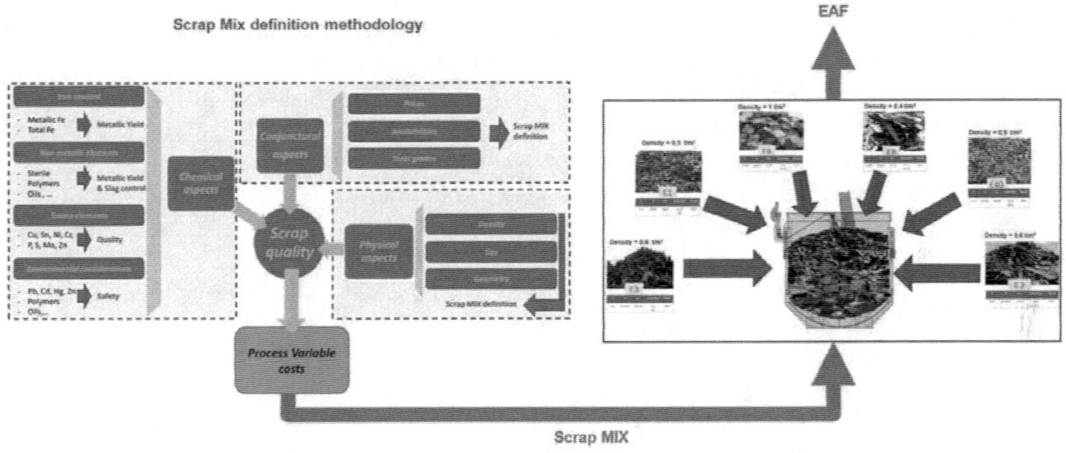

Figure 4, Scrap mix definition strategy based on TCO concept [5]

The Value In Use (VIU) has proven to be a very useful criteria to relate scrap quality with the operating cost of the Electric Arc Furnace operations. It may be used as the variable to be minimized when optimizing the global scrap mix TCO by taking into consideration all the physical, chemical constrains (related to steel facility and products) as well as the scrap market situation and production forecast as depicted in Figure 4. Once the optimal scrap mix is established for a production period, the ferrous scrap reposition strategies can be systematically evaluated over time for the scrap yard as proposed in Figure 5.

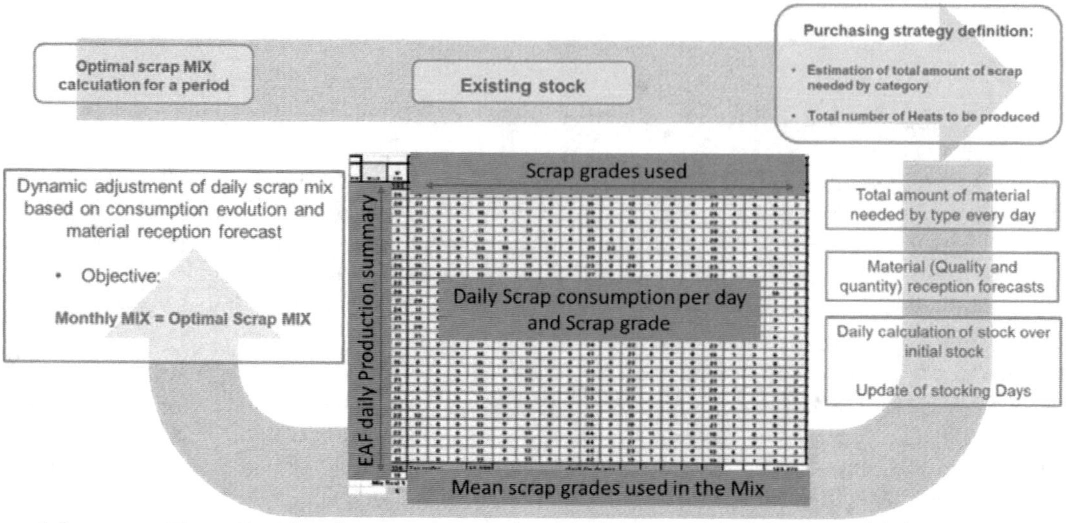

Figure 5, Scrap reposition methodology by comparing calculated Optimal Scrap Mix definition and daily consumption follow up

Although the methodology proposed in Figure 5 offers a highly technical solution to optimize purchasing, reposition and consumption decisions of ferrous materials in the EAF scrap yard facilities, it also presents an important drawback which is linked to the huge amount of accurate information that is required for launching the overall scrap mix optimization methodology.
Additionally, there are usually some events that can destroy the overall optimization and require punctual adjustment in the scrap yard management:

- Deliveries of scrap at the plant do not always arrive as scheduled
- The quality of the materials received at the plant do not match the estimated qualities when the mix optimizations were launched.
- The existing stocks do not match with initial estimations
- New market opportunities arise that did not exist when the mix optimization processes began
- Production scheduling usually undergo adjustments due to production incidents or reprioritization of customer demands

For all these cases, together with scrap VIU, Cu content of scrap is the most impacting factor in the scrap material assessment when punctually correcting the scrap mix. This is the reason of proposing a new methodology for estimating the influence of Cu content of Ferrous Scrap materials in the Total Cost Of Operation (TCO) of scrap mix in Electric Arc Furnace steelmaking.

3. The need for this work

The objective of this work is to provide an easy-to-use tool to facilitate the decision making when choosing between scraps with different copper content and price and considering the copper limit of the steel to be produced.
Given the input parameters, the tool provides the TCO of the optimum mix and shows a graph to illustrate how the TCO would vary with the copper content of the scrap. Running the calculation again with another set of parameters, will again provide the TCO of the optimum mix, allowing for comparison and decision making regarding the scrap to be purchased.
The result of this tool must be considered together with the VIU of the scrap and any other relevant aspects.

4. Parametric study

The TCO provides information of all costs associated to melting a scrap mix in the EAF. Starting from data related to the scrap types available for purchase, their estimated chemical composition, price, process related restrictions and other internal considerations, the optimum mix is calculated and its TCO is calculated, and the purchase strategy is defined. Based on this calculation methodology, the present

work has fixed most of the parameters to analyse the impact of those that were left open to change, namely, the copper content of the scrap types available, their price and the copper content in the steel products. This work has been carried out for two types of steel: one with high copper content requirements and one with lower copper content requirements.
Considering scrap types with different contents of copper and price as well as a range of copper allowed in the mix, a parametric study has been performed for low quality steel (greater quantity of copper allowed in the final product) and high-quality steel (more onerous restriction related to copper content in the final product).
For each scenario, only two scrap types were allowed: a good quality scrap (low residual scrap in this paper) and a lower quality scrap (high residual scrap in this paper).
The values considered for each parameter were as follows:
Parameters considered were:

1. Cu content of the low residual scrap, $\%Cu_{HR}$
2. Cu content of the high residual scrap, $\%Cu_{LR}$
3. Price gap between the low and high residual scrap, ΔP, where $\Delta P = P_{LR} - P_{HR}$
4. Total Cu content allowed in the mix, [%Cu]

For all the scenarios generated, the TCO of the optimum mix was calculated to establish relationships with the copper content of each scrap type.
Due to the typical differences between the more and less restrictive steel grades, the following values of the parameters described above were considered as shown in Table 1.

Table 1. Range of parameters considered in this work.

	Less restrictive steel grades (typically long)	More restrictive steel grades (typically flat)
%CuLR	0.06 – 0.08 – 0.10 – 0.12 – 0.14	Ignored
%CuHR	0.05 – 0.10 – 0.15 – 0.20 – 0.25 – 0.30 – 0.35 – 0.40 – 0.45 – 0.50 – 0.55	0.01 – 0.05 – 0.10 – 0.15 – 0.20 – 0.25 – 0.30 – 0.35 – 0.40 – 0.45 – 0.50 – 0.55
[%Cu]	0.1 – 0.2 – 0.3 – 0.4 – 0.5 – 0.6	0.01 – 0.02 – 0.03 – 0.04 – 0.05 – 0.06 – 0.07 – 0.08 – 0.09 – 0.1
DP (€/t)	10 – 20 – 30 – 40 – 50	10 – 20 – 30 – 40 – 50
Total # of scenarios	1650	600

The choice of prices for the low and high residual scraps has been made based on historic datasets of typical scraps. However, the methodology developed in this work should not be dependent on specific prices, as these are volatile and may and will change over time and circumstances.
For this reason, all data obtained in the parametric study have been normalised with TCO_{REF}, and the normalised TCO value is called $\widehat{TCO}$ or TCO_{NORM}

$$\widehat{\boldsymbol{TCO}} = \frac{\boldsymbol{TCO}}{\boldsymbol{TCO}_{\mathrm{REF}}} \tag{1}$$

Figure 6 shows an example of the result of this normalisation for [%Cu] = 0.3 for the case of less restrictive steel grades.

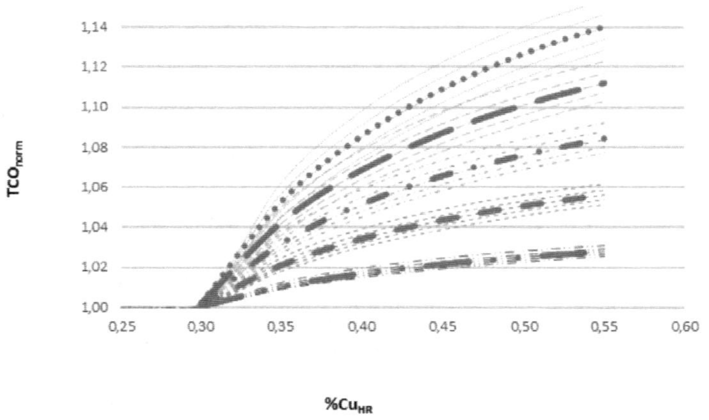

Figure 6, [%Cu] = 0.3 ; ΔP = 10 - 50€/t ; %CuLR = 0.06 – 0.14

The thick red lines represent the average values for all $\%Cu_{LR}$ for each value of ΔP, with ΔP = 10€/t being the one at the bottom and ΔP = 50€/t being the one at the top.

Some preliminary observations from the simulations are as follows:

1. The minimum TCO is obtained when a mix is fully composed of high residual scrap. This value of the TCO will be called TCO_{REF}.
2. The higher the value of [%Cu] the more high-residual scrap is used. As the maximum Cu content in the scrap has been considered as 0.55, for [%Cu]=0.6, all scrap is high residual, and therefore the TCO is the lowest and equal to TCO_{REF}.
3. The lower the value of %Cu in the high residual scrap, the more this type of scrap will be used, and therefore, the TCO will also be lower.
4. For values of the %Cu in the high residual scrap up to the value of [%Cu], the mix is composed of high residual scrap only, and therefore, the TCO will have its minimum possible value, TCO_{REF}.

5. Analysis of results

The variability of scrap price and Cu content implies that a general assessment method was needed, regardless of the surrounding circumstances. The results were analysed in terms of the Cu content of the low residual scrap, maximum Cu content of the mix and the price difference between the low and high residual scraps, as discussed in the following sections.

5.1. Low residual Cu content

Five levels of Cu content were considered in this analysis. Figure 7 shows the curves for [%Cu]=0.3, for different levels of Cu content in the low residual scrap. Each group formed by 5 curves represents the 5 levels of Cu content of low residual scrap for a given price difference, ΔP. The red line in the middle of each group represents the mean curve for each group. The average curve fits accurately in all cases with the middle value of Cu content of low-residual scrap, i.e. 0.10.

The error of considering the average curve as opposed to the actual curve for each value of $\%Cu_{LR}$ was calculated as:

$$\boldsymbol{E = 100 \times \left|\frac{\widehat{TCO}_{Real} - \widehat{TCO}_{Avg}}{\widehat{TCO}_{Real}}\right|} \tag{2}$$

Where:

$\widehat{TCO}_{Real}$	is the real calculated value of the normalised TCO
$\widehat{TCO}_{Avg}$	is the average value of the normalised TCO for a given ΔP
E	is the error of using the average TCO as opposed to the calculated TCO

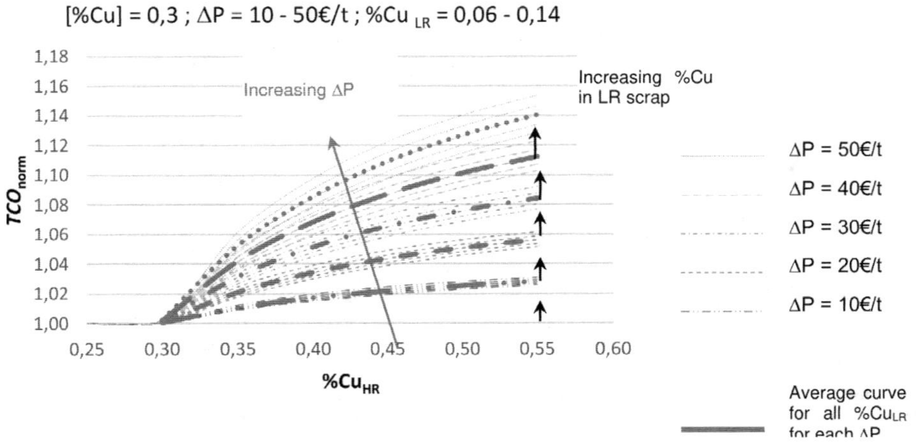

Figure 7, Typical curves of TCO vs. %Cu in High Residual scrap

Table 2 shows the maximum error that occurs when the average curve is considered as opposed to the actual real value for the relevant value of Cu content in the low residual scrap

Table 2. Maximum error when considering the average curve for each ΔP and [%Cu]..

	Maximum error between average value and maximum or minimum value for varying low residual Cu content (%).					
[%Cu] ΔP	0.1	0.2	0.3	0.4	0.5	0.6
10€/t	0.39	0.56	0.26	0.16	0.05	0.00
20€/t	0.75	1.10	0.51	0.31	0.11	0.00
30€/t	1.10	1.62	0.75	0.45	0.16	0.00
40€/t	1.42	2.10	0.98	0.59	0.21	0.00
50€/t	1.73	2.57	1.20	0.73	0.27	0.00

The summary table shows a maximum error of 2.57% when considering the average curve. With the aim of finding a simple way of considering the influence of copper content in scrap, this error was ignored, and the average curve was used in what followed.

5.2. Maximum Cu content in steel mix, [%Cu]

Figure 8 shows the average curves of TCO_{NORM} for each value of ΔP and [%Cu].

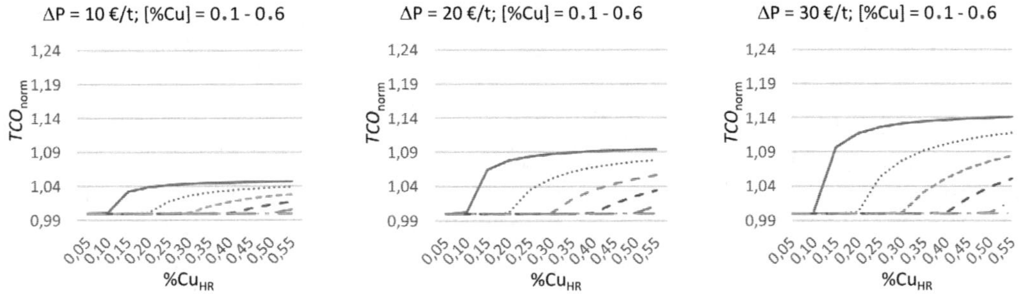

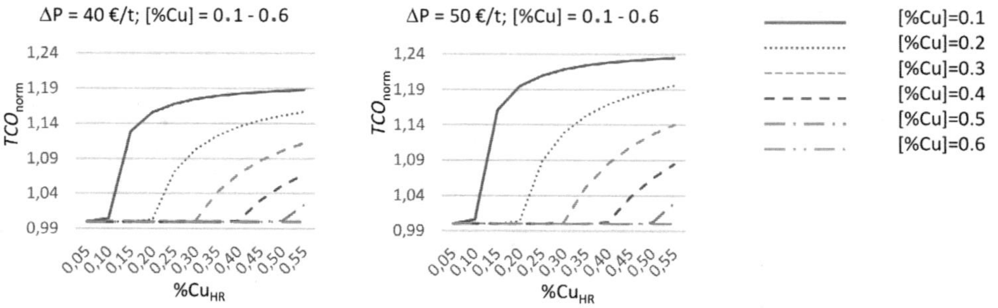

Figure 8, Average curves for the different values of low residual scrap for [%Cu] from 0.1 to 0.6

These graphs show that the higher the amount of copper allowed in the mix is, the higher is the amount of high residual scrap, and therefore the lower the cost. For values of $\%Cu_{HR}$ < [%Cu] TCO_{NORM} is equal to 1, as 100% of the mix is composed of high residual scrap, and therefore $TCO = TCO_{REF}$.

5.3. Price difference between high and low residual scrap, ΔP

Figure 9 presents the average curves of the TCO in terms of the $\%Cu_{HR}$ for each value of [%Cu] and ΔP.

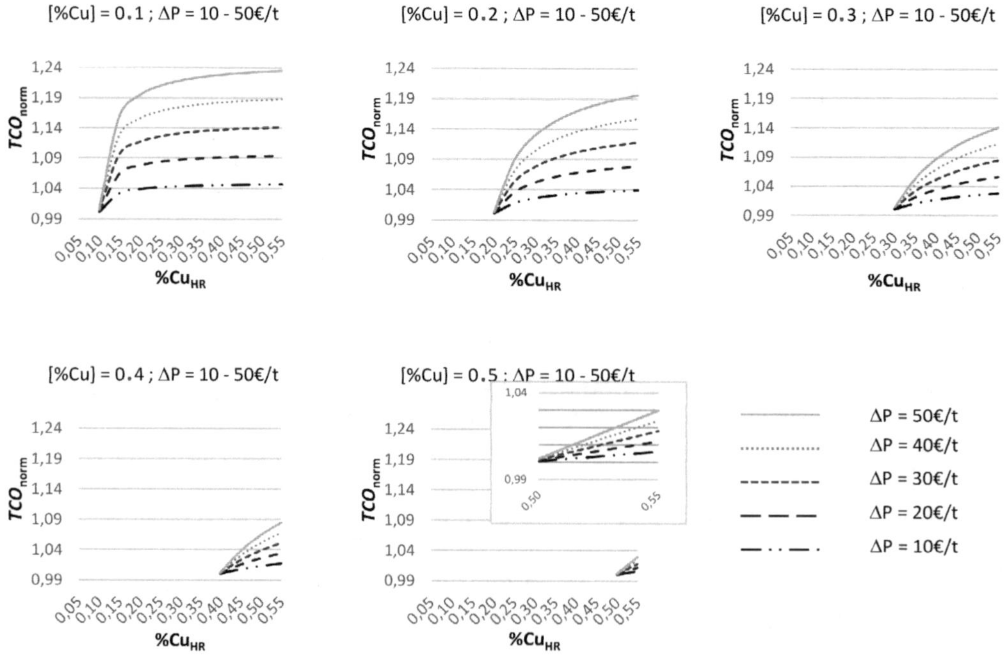

Figure 9, Average curves of the TCO in terms of the %CuHR for each value of [%Cu] and ΔP

The curves in each graph show that a higher TCO_{NORM} is obtained when there is a higher price difference between the two scrap types, and this is more acute when a lower copper content of the mix is allowed. For the case of the more restrictive steel type, Figure 10 shows the curves obtained for TCO_{NORM} in terms of $\%Cu_{HR}$ for different values of ΔP, grouped by [%Cu].

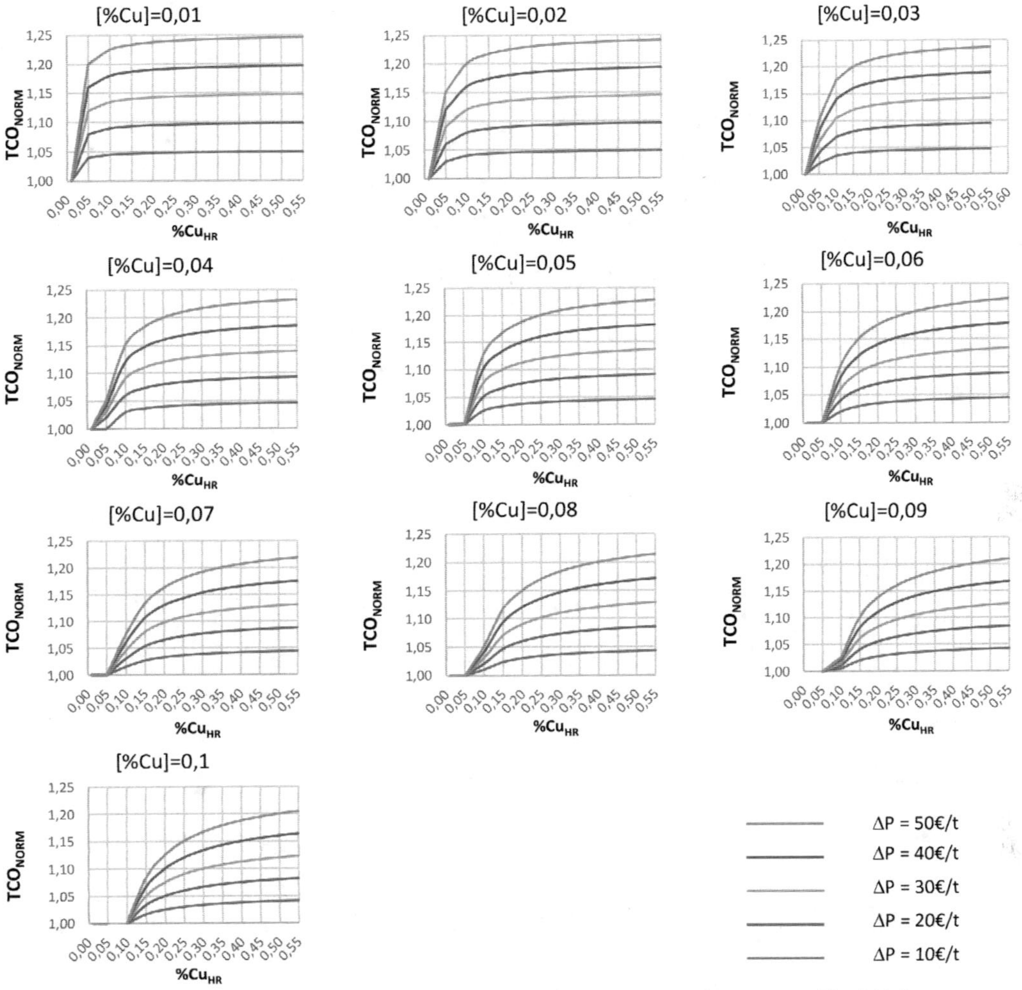

Figure 10, TCONORM in terms of %CuHR for different values of ΔP, grouped by [%Cu]

5.4. Equations relating cu and other parameters

To develop the relationships that can help establish the impact of the copper content in scrap on the TCO of the heat, the starting point is the set of curves for a given [%Cu] value and different price differences. This is because $\widetilde{TCO}$ is equal to 1.0 up to just before the point where the copper content in the high residual scrap is equal to [%Cu]. This provides a common starting point for all the curves for a given value of [%Cu].

The equations for the different cases have been derived following the procedure described below for the case of [%Cu] = 0.4.

Figure 11 shows the average curves for TCO_{NORM} for different values of ΔP and [%Cu] = 0.4. It also presents the 2nd degree polynomic expressions that fit the curves.

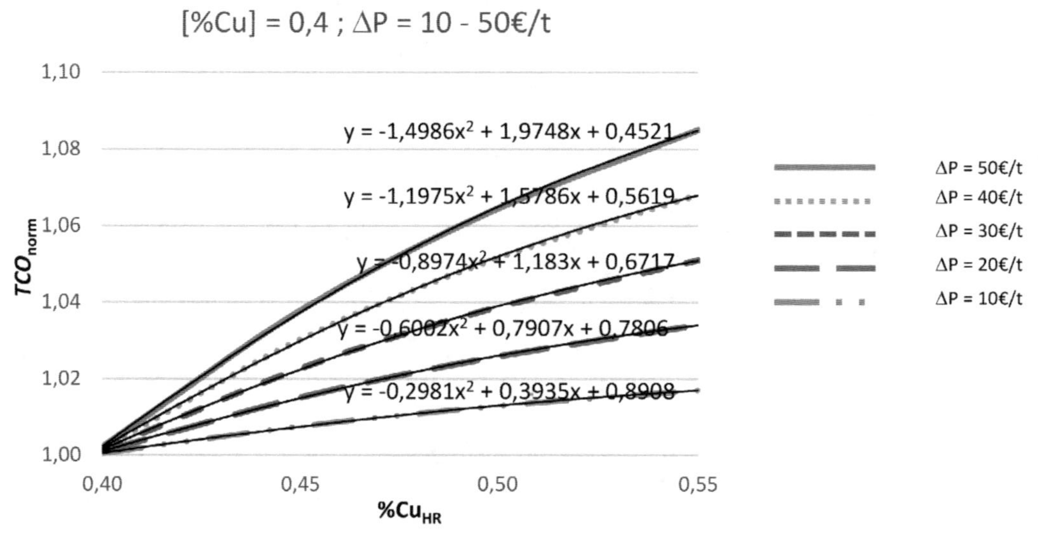

Figure 11, Average curves for TCONORM for different values of ΔP and [%Cu] = 0.4.

The relationship between $\widehat{TCO}$ and %Cu_{HR} can be defined as a parabolic equation of the shape:

$$\widehat{TCO} = ACu_{HR}{}^2 + B\ \%Cu_{HR} + C \quad (3)$$

where A, B and C are as given in Table 3

Table 3. Coefficients A, B and C for [%Cu] = 0.4.

ΔP (€/t)	B	C	D
10	-0.2981	0.3935	0.8908
20	-0.6002	0.7907	0.7806
30	-0.8974	1.183	0.6717
40	-1.1975	1.5786	0.5619
50	-1.4986	1.9748	0.4521

These values of the constants B, C and D that define the value of $\widehat{TCO}$ for a given value of ΔP, is shown in Figure 12.

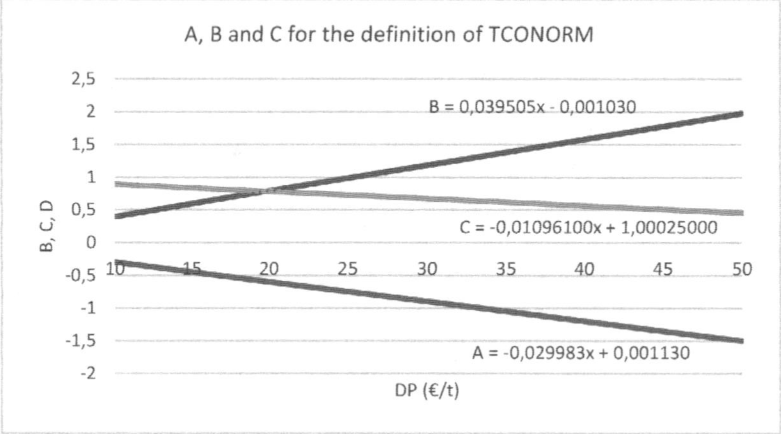

Figure 12, Relationship between A, B and C with ΔP

It is observed that the coefficients of the polynomic expressions follow a linear trend that can be expressed in terms of ΔP as:

$$A, B, C = a\Delta P + b \tag{4}$$

Where a and b are given in Table 4.

Table 4. Coefficients defining A, B and C in terms of ΔP.

	a	b
A	-0.029983	0.001130
B	0.039505	-0.001030
C	-0.010961	1.00025

Therefore, TCO_{NORM} can be expressed in terms of ΔP and $\%Cu_{HR}$ and a set of coefficients that depend on [%Cu] alone.

For the less restrictive steel types, up to 3rd degree polynomic expressions were obtained, thus leading to an equation as follows:

$$TCO_{NORM} = (a_1\Delta P+a_2) \times \%Cu_{HR}^3 + (b_1\Delta P+b_2) \times \%Cu_{HR}^2 + (c_1\Delta P+c_2) \times \%Cu_{HR} + (d_1\Delta P+d_2) \tag{5}$$

Following the same procedure, the coefficients to be used for all scenarios analyzed can be seen in Table 5.

Table 5. Coefficients for less [%Cu] restrictive steel grades.

[%Cu]	%CuHR	a1	a2	b1	b2	c1	c2	d1	d2
0.1	≤[%Cu]	0	0	0	0	0	0	0	1
	≤0.15	0	0	0	0	0.0619	0.0003	-0.0061	1
	>0.15	0.0553	0.001	-0.0713	-0.0009	0.31	0.0003	0.00002	1
0.2	≤[%Cu]	0	0	0	0	0	0	0	1
	>[%Cu]	0.17695	-0.02377	-0.2398	0.0001	0.111	-0.0002	-0.0139	1
0.3	≤[%Cu]	0	0	0	0	0	0	0	1
	>[%Cu]	0	0	-0.0349	-0.0006	0.0404	0.0005	-0.0089	0.9999
0.4	≤[%Cu]	0	0	0	0	0	0	0	1
	>[%Cu]	0	0	-0.02998	0.00113	0.039505	-0.00103	-0.010961	1.00025
0.5	≤[%Cu]	0	0	0	0	0	0	0	1
	>[%Cu]	0	0	0	0	0.0111	0.00009	-0.0055	1
0.6	≤[%Cu]	0	0	0	0	0	0	0	1
	>[%Cu]	Not enough information was available to generate this information (one single point where $TCO_{NORM} \neq 1$)							

Table 6 provides the coefficients obtained for the more restrictive steel types, where up to 5th degree polynomic expressions were obtained, thus leading to an equation as follows:

$$TCO_{NORM} = (a_1\Delta P+a_2) \times \%Cu_{HR}^5 + (b_1\Delta P+b_2) \times \%Cu_{HR}^4 + (c_1\Delta P+c_2) \times \%Cu_{HR}^3 + (d_1\Delta P+d_2) \times \%Cu_{HR}^2 + (e_1\Delta P+e_2) \times \%Cu_{HR} + (f_1\Delta P+f_2) \tag{5}$$

Table 6. Coefficients for more [%Cu] restrictive steel grades.

[%Cu]	**$\%Cu_{HR}$**	**a1**	**a2**	**b1**	**b2**	**c1**	**c2**	**d1**	**d2**
0.01	≤0.01	0	0	0	0	0	0	0	0
	≤0.05	0	0	0	0	0	0	0	0
	>0.05	0	0	0	0	0.0252	-0.00002	-0.0285	0.00003
0.02	≤0.01	0	0	0	0	0	0	0	0
	≤0.05	0	0	0	0	0	0	0	0
	>0.05	0	0	-0.2101	-0.0039	0.3024	0.008	-0.1572	-0.0044
0.03	≤0.01	0	0	0	0	0	0	0	0
	≤0.10	0	0	0	0	0	0	-0.2295	0.0069
	>0.10	0	0	0	0	0.0299	0.0006	-0.0375	-0.0005
0.04	≤0.01	0	0	0	0	0	0	0	0
	≤0.05	0	0	0	0	0	0	0	0
	>0.05	1.3844	12.662	-2.4408	-21.276	1.6625	13.598	-0.5529	-4.0083

0.05	≤0.01	0	0	0	0	0	0	0	0
	≤0.05	0	0	0	0	0	0	0	0
	>0.05	2.1155	0.3969	-3.6979	-0.68	2.4917	0.4378	-0.813	-0.1309
0.06	≤0.01	0	0	0	0	0	0	0	0
	≤0.05	0	0	0	0	0	0	0	0
	>0.05	1.0674	0.0279	-1.9631	-0.0563	1.4168	0.0378	-0.5087	-0.01
0.07	≤0.01	0	0	0	0	0	0	0	0
	≤0.05	0	0	0	0	0	0	0	0
	>0.05	0	0	-0.1925	0.026	0.3158	-0.0298	-0.1951	0.011
0.08	≤0.01	0	0	0	0	0	0	0	0
	≤0.05	0	0	0	0	0	0	0	0
	>0.05	-1.1393	-0.013	1.692	-0.0117	-0.8509	0.0256	0.135	-0.0108
0.09	≤0.05	0	0	0	0	0	0	0	0
	≤0.10	0	0	0	0	0	0	0	0
	>0.10	1.08	0.153	-2.069	-0.2971	1.5691	0.206	-0.5981	-0.0639
0.1	≤0.10	0	0	0	0	0	0	0	0
	>0.10	0	0	-0.3492	-0.0135	0.5536	0.0132	-0.3282	-0.0037

Table 6. Coefficients for more [%Cu] restrictive steel grades *(continued)*

[%Cu]	$\%Cu_{HR}$	e1	e2	f1	f2
0.01	≤0.01	0	0	0	1
	≤0.05	0.0994	0.00004	-0.001	1
	>0.05	0.0104	0.00005	0.0036	1
0.02	≤0.01	0	0	0	1
	≤0.05	0.0753	-0.0006	-0.0008	1
	>0.05	0.0357	0.0008	0.0016	1
0.03	≤0.01	0	0	0	1
	≤0.10	0.0643	-0.0011	-0.0006	1
	>0.10	0.016	0.00007	0.0023	1
0.04	≤0.01	0	0	0	1
	≤0.05	0.0307	-0.2053	-0.0003	1.0021
	>0.05	0.0924	0.5461	-0.0022	0.9732
0.05	≤0.01	0	0	0	1
	≤0.05	0.00E+00	3.85E-02	-2.00E-17	9.96E-01
	>0.05	0.1321	0.0173	-0.0048	0.9992
0.06	≤0.01	0	0	0	1
	≤0.05	0	0	0	1
	>0.05	0.0949	0.0008	-0.0036	1
0.07	≤0.01	0	0	0	1
	≤0.05	0	0	0	1
	>0.05	0.056	-0.0012	-0.0024	1.0001
0.08	≤0.01	0	0	0	1
	≤0.05	0	0	0	1
	>0.05	0.0155	0.0014	-0.0011	0.9999
0.09	≤0.05	0	0	0	1
	≤0.10	1.06E-02	-5.00E-04	-5.00E-04	1.00E+00
	>0.10	0.1198	0.0086	-0.0068	0.9996
0.1	≤0.10	0	0	0	1
	>0.10	0.0898	0.0004	-0.0062	0.9999

6. Model validation

The validation has been carried out based on real data from several ArcelorMittal sites. From each of them, process information for heats, data of the mix as well as the copper content of each scrap type were collected and used for validation. In total, 20 different scenarios for long steel products and other 20 scenarios for flat steel products were simulated for validating the model. The scrap with the most copper content was selected as the high residual one and the rest were grouped together to form the

low residual scrap, whose copper content and price was calculated as the weighted average of the scraps composing it.
Validation cases include data for less restrictive steel qualities in terms of copper content, as well as data for more onerous conditions in terms of copper content allowed.
The difference between the TCO calculation by using the internal tool that uses all the parameters and the TCO estimation by using the expressions presented in this paper were in all cases below 2.5%, calculated as the absolute value of $100x(TCO_{calc}-TCO_{est})/TCO_{calc}$.
For the case where a higher amount of copper is allowed in the final mix, validations have been carried out using heats of EAFs. Two price differences have been considered, for 5 different values of copper content in the high residual scrap, 5 different values of copper content in the mix and 2 different values of copper content in the low residual scrap. The results can be seen in the graphs in Figure 13.

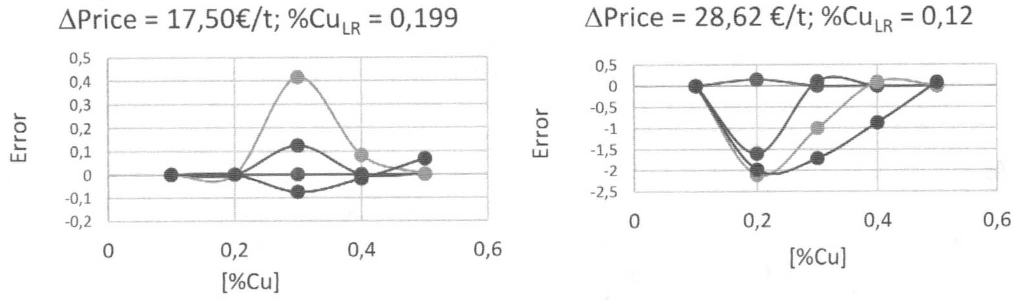

Figure 13, Error calculated in the validation of the methodology for higher content of copper in the mix

For the case of the more restrictive copper content in the mix, BOF heats were selected and the calculation model as well as the full TCO calculation model were run for 4 levels of copper contents in the scrap under consideration, 19 values of copper content allowed in the mix and 2 different price differences. The errors obtained for all of them are shown in figure 14.

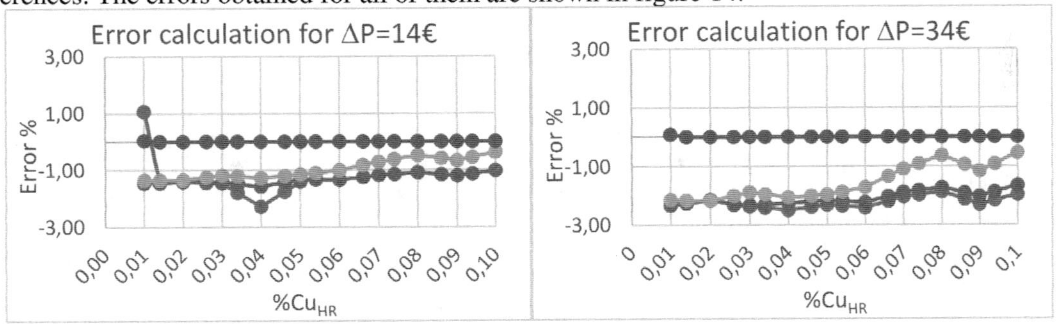

Figure 14, Error calculated in the validation of the methodology for low content of copper in the mix

7. Conclusions

Until now, two options were available to steelmakers' purchasing departments when it came to deciding on the strategy in terms of scrap quality.
The first option is based on a highly complex calculation in existing models, taking into account many parameters and information related to the scrap, thermodynamics, process, client requirements, market situation and changing conditions, that would only allow for medium-term decisions. The input information required is such, that this model cannot be run daily, and therefore the strategical decisions on the scrap purchase must be maintained for a few weeks until a new scenario can be run.
The second option would be to systematize Cu scrap-based periodic optimizations, using the initial complex approach for purchasing strategy corrections based on the copper needs alone, leaving to one side other relevant factors that could be relevant.

The development of this mathematical model together with the traditional VIU allows for a fast assessment of the optimum choice of scrap, without ignoring other technical aspects that are relevant for steelmakers.

8. Aknowledments

Partial financial support of this work by the Basque Government (Hazitek ALLIRON ZE-2019/00005 – Development of strategic Steel grades to answer ALL the requirements of advanced customers to IRONclad the ArcelorMittal Sestao CSP) is gratefully acknowledged.

9. REFERENCES

[1] https://worldsteel.org/wp-content/uploads/World-Steel-in-Figures-2022.pdf
[2] https://corporate-media.arcelormittal.com/media/bg1n545s/arcelor-mittal-integrated-annual-review-2021.pdf
[3] WorldSteelAssociation. Global steel industry: outlook, challenges and opportunities. 5th International Steel Industry & Sector Relations Conference. April 20th, 2017 – Istanbul
[4] J.-P. Birat, X. Le Coq, P. Russo, E. Gonzales, J.J. Laraudogoitia, Quality of heavy market scrap: development of new and simple methods for quality assessment and quality improvement. (2002 - RFCS)
[5] Asier Vicente Rojo. New methods for ferrous raw materials characterization in electric steelmaking. PhD thesis (2020)
[6] Factors Influencing Total Energy Consumption In Arc Furnaces. Graphtech. Phoenix, Arizona, : s.n., 2001. 59th Electric furnace conference.
[7] ArcelorMittal. Steelmaking: EAF (TCO & Quality training). s.l. : ArcelorMittal University, 2010.

IOP Conf. Series: Materials Science and Engineering 1309 (2024) 012007 doi:10.1088/1757-899X/1309/1/012007

Laboratory Scale Evaluation of the Slag Foaming Behavior

Andreas Pfeiffer[1,4], Kathrin Thiele[2], Gerald Wimmer[1], Johannes Schenk[3]

[1] Primetals Technologies Austria GmbH., Linz, Austria
[2] Christian Doppler Laboratory for Inclusion Metallurgy in Advanced Steelmaking, Montanuniversität Leoben, Austria
[3] Chair of Ferrous Metallurgy, Montanuniversität Leoben, Austria
[4] Author to whom any correspondence should be addressed. andreas.pfeiffer@primetals.com

Abstract. Due to the ambitious climate targets of the European Union, one can expect that the electric arc furnace (EAF) will gain greater importance in the future of steelmaking. Since slag foaming is a decisive factor in an efficient process, understanding this phenomenon is essential when applying hydrogen-based direct reduced iron (DRI). Therefore, a method was developed to check different slag compositions concerning their foaming behavior. Slag samples are melted, and a carbon carrier is added. After a selected reaction time, the crucible is quenched in liquid nitrogen, superficially freezing the state while foaming. Afterward, it is halved, providing metallographic examination and height measurement possibilities. Three slags were tested, MgO-saturated EAF slag, MgO-unsaturated EAF slag, and electrical Smelter-like slag. Digital and scanning electron microscopy with energy dispersive x-ray spectroscopy are used to compare the slags and evaluate the method. The Smelter slag shows no foamability, unaffected by the FeO content. Contrary, good foamability can be observed for EAF slags.

1. INTRODUCTION

Slag foaming happens during steelmaking in basic oxygen furnaces (BOF) and electric arc furnaces (EAF). While in the former, it occurs during the period of maximum decarburization, in the latter, it improves the process in multiple ways. The foam covers the arc, increasing the efficiency from approximately 40 to 60 or even 90% [1]. Further, it influences interactions with the atmosphere, e.g., nitrogen pickup [2], and the reduced heat radiation protects the refractory materials [3].

Two preconditions must be given for a foaming slag in the EAF, the formation of gas bubbles and specific slag properties [4]. The former is usually achieved via the decarburization reactions, e.g., in 1, with solid carbon being dominant in the EAF and dissolved carbon in the BOF. The highly endothermal nature, with $\Delta h=+124.9$ kJ/mol (calculated by FactSage™ 8.3, 1600°C), characterizes the reaction in equation (1).

$$(FeO)+[C] \text{ or } C_{sol} \Leftrightarrow [Fe] + \{CO\} \quad (1)$$

The demanded slag properties are usually defined by surface tension, density, viscosity, and the number of solid particles, with the latter two being the most essential. Based on laboratory tests, these parameters can be linked to the so-called slag foaming index, a concept described by Ito and Fruehan [5,6]. However, most experimental investigations of the slag foaming behavior are performed with either inert gas [5–11] or cold models [12–14]. As the endothermal reaction 1 may itself influence the slag foaming behavior and there is only a little research published in this case, e.g., by Echterhof and Pfeifer [15], it is worth a closer look. Therefore, a new method is developed to study reaction-induced slag foaming.

2. MATERIALS AND METHODS

The tests were performed in an Indutherm MU900 induction furnace (Indutherm Erwärmungsanlagen GmbH., Walzbachtal, Germany). Figure 1 shows the experimental setup. The synthetic slag mixture is charged in an alumina crucible with a piece of ultra-low carbon (ULC) steel to stabilize Fe^{2+}. The refractory crucible is placed in the graphite crucible of the furnace and indirectly heated to 1600 °C, measured with the furnace thermocouple. The chamber is flushed with 1000 Nl/h nitrogen. After reaching the target temperature, the sample is slightly stirred to homogenize it and kept at a constant temperature for 15 min. Then calcined petrol coke (CPC) with C_{fix}~ 99 wt.-% and a particle size between 0.5 and 1 mm is added and mixed using a Mo wire. The amount is chosen equimolar according to reaction 1. After reaching the demanded foaming time, the refractory crucible is extracted and quenched in liquid nitrogen to freeze the foam morphology.

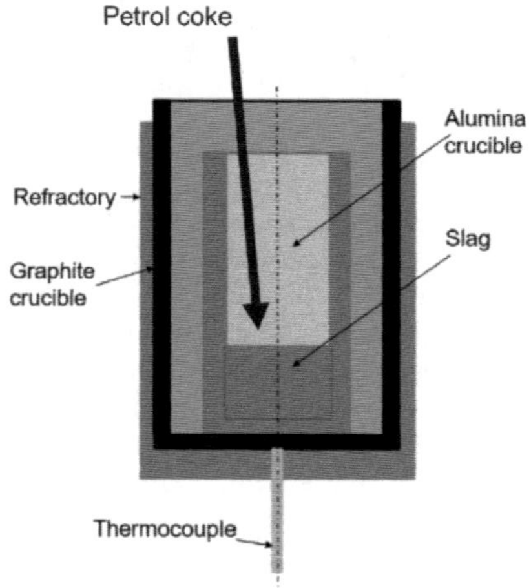

Figure 1. Experimental setup

The crucibles are fixed with fabric tape to avoid breakage from thermal stresses and filled with a low-viscosity embedding mass. That enables cutting and metallographic examination of the cross-section.

A Keyence VHX 7000 (Keyence Corporation, Osaka, Japan) digital optical microscope, and a field-emitting scanning electron microscope (JEOL 7200F, JEOL Germany GmbH. Freising, Germany) equipped with a 100 mm² SDD-EDS detector (Oxford Instruments Ul-tim Max 1000; Oxford Instruments GmbH NanoAnalysis, Wiesbaden, Germany) allowed for evaluation of the samples. The height after foaming was measured using the digital microscope, with the volume gain being calculated based on the initial slag volume. The latter is approximated from the slag density according to the approach described by Heo et al. [16]. Equation (2) standardizes the determined values.

$$\frac{\Delta V}{V_0} = \frac{V_1 - V_0}{V_0} \tag{2}$$

Table 1 summarizes the four slag compositions analyzed in this test series. Smelter-1 and Smelter-2 represent slags in electrical Smelters with two different FeO contents. EAF and EAF-sat are typical for EAF slags, with the former being fully liquid and the latter saturated with MgO.

Table 1. Target slag compositions; Liquidus temperature acc. to FactSage™ 8.2; *Fe assumed to be Fe^{2+}

wt.-%	Smelter-1	Smelter-2	EAF-sat	EAF
CaO	43.9	41.9	31.6	35.0
SiO_2	39.6	38.0	19.4	22.0

Al_2O_3	4.89	4.90	9.10	8.00
MgO	6.00	5.70	11.9	9.00
FeO*	5.62	9.50	28.0	26.0
B2 / 1	1.1	1.1	1.6	1.6
$T_{Liquidus}$ / °C	1399	1381	1681	1529

Slag preparation went as follows. An iron oxide-free premixture was made and premelted in the induction furnace directly in the graphite crucible. That premelt was quenched by casting it on a steel plate. The premelt was mixed with iron oxide for the actual tests and finely adjusted with oxide powders according to the target composition. CaO, MgO, SiO2, and Al2O3 were applied in the form of technically pure oxides with 95-99 % pureness; FeO, however, was added as prereduced hematitic iron ore. This procedure provides two advantages. On the one hand, the volume is reduced compared to pure oxides, and on the other hand, a low melting liquid pool forms during heating. A more uniform melting process can be expected from that. In total, 10 tests were performed with foaming times between 0 and 5 min.

3. RESULTS

Figure 2 shows digital microscope images of an EAF-sat, EAF, and the Smelter-2 test. In A), the foam exceeds the crucible, indicative of outstanding foamability. The center is hollow. It is not completely clear if this is also the case during the tests. One can expect the morphology during the test to have a foamy structure throughout the whole crucible cross-section. Although similar at first sight, two differences can be noticed in the unsaturated slag in B). Firstly, a slight circumferent depression at the top can be seen, indicative of lower foam stability. Secondly, the morphology consists of one big and many tiny blisters. However, it seems absent from blisters approximately 3-5 mm in size, which are common in A). As the thermal conductivity of the crucible is limited, the slag cannot be quenched fast enough in the center, preventing the formation of a stable solid structure. The Smelter-slag in C) shows a dense morphology with one blowhole in the center. This hole results from the top being solidified before the crucible center.

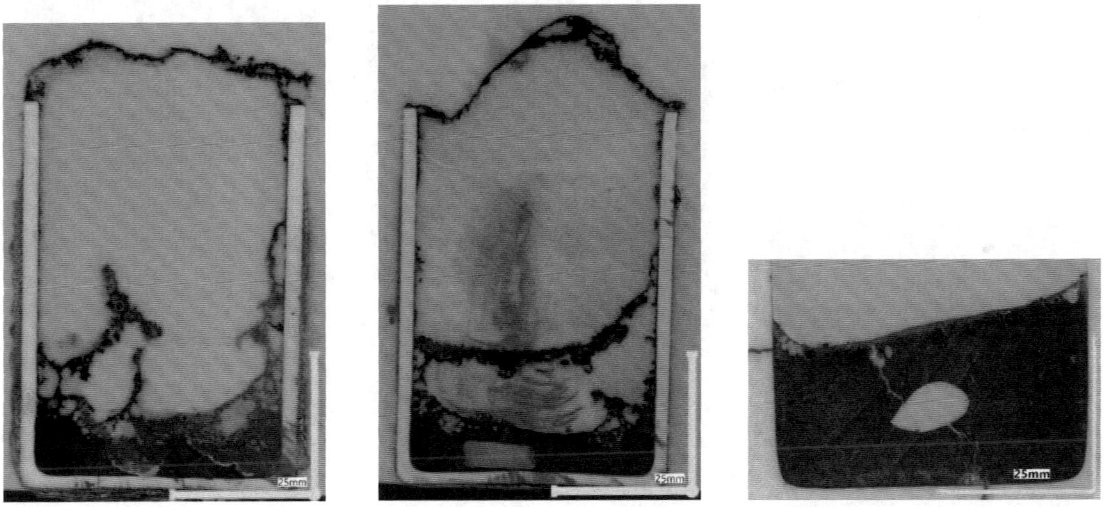

A) B) C)

Figure 2. Digital microscope images of A) EAF-sat; B) EAF; C) Smelter-2

Figure 3 shows the specific volume increases from equation 2 as a function of foaming time. The test with foaming time 0 s represents a reference test without adding CPC. Looking at the EAF and EAF-sat points, it can be observed that no significant difference occurs. In this case, it must be mentioned that the slag in test EAF-sat with 2 min foaming time slightly collapsed. That leads to a reduced value. The dots for the reference (t=0 min) and the Smelter tests do not show a significant volume increase. The fact that they are not at 0 can be explained in two ways: a) the blowholes increase the volume, and b) the model-based density calculation may not be totally accurate.

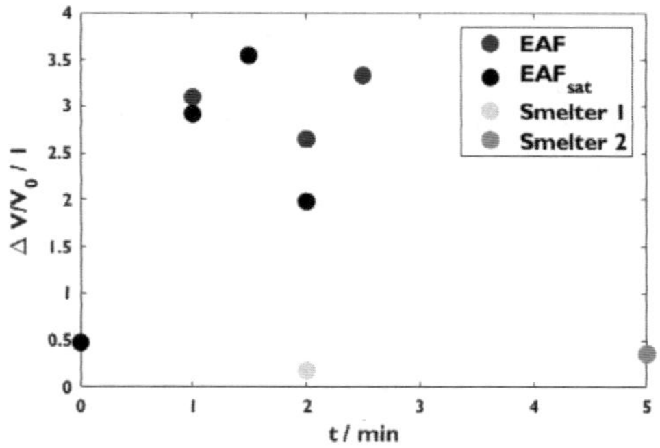

Figure 3. Specific volume increase vs. foaming time of the tests

Figure 4 gives the SEM image as well as the O and Fe distributions of an EAF-sat test sample. The metallic iron proves the reduction reaction between CPC and (FeO). The fact that metallic iron is mainly found in the foam indicates pronounced wetting between metal droplets and slags, preventing them from moving to the crucible bottom.

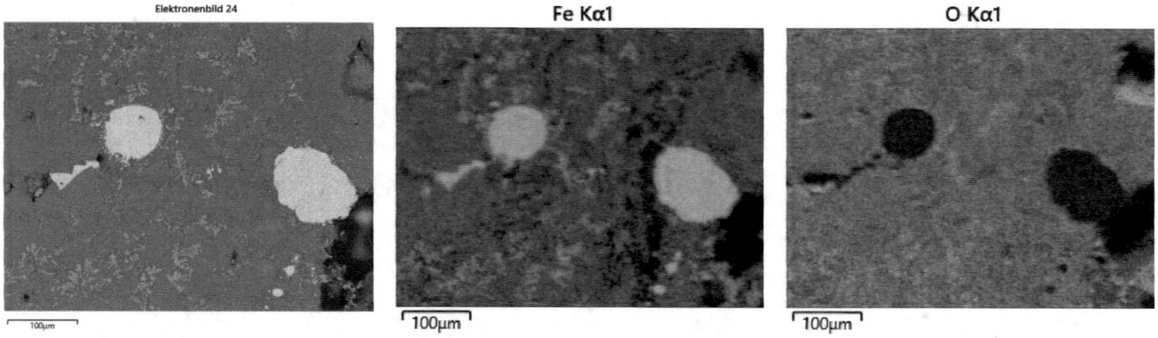

Figure 4. SEM image, O and Fe EDX-distribution of EAF-sat slag in the foam

Figure 5 visualizes SEM and EDX images from the same sample but in the lower part of the crucible. The structure appears infused with dendrites, rich in Fe and Mg, indicative of Magnesiowustite—nevertheless, the dendritic morphology results from a limited cooling rate which correlates with the observations from above.

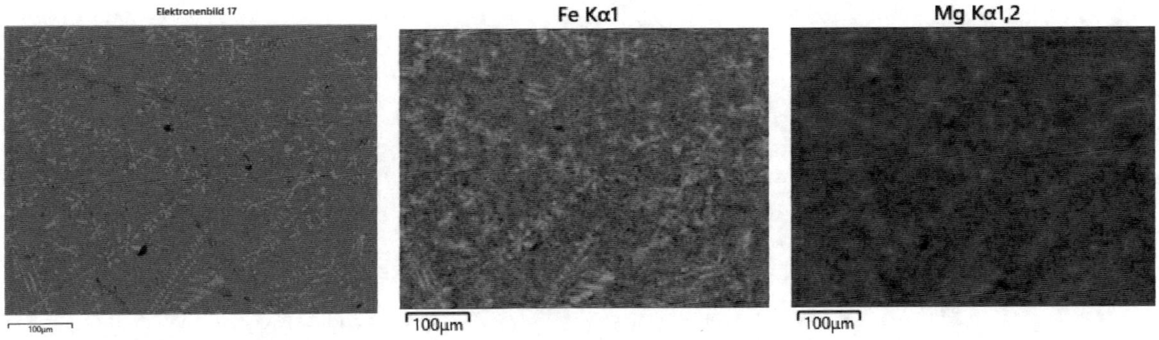

Figure 5. SEM image, Mg and Fe EDX-distribution of EAF-sat slag in the crucible center

Figure 6 illustrates the surface-near area. The enrichment with Al and Mg indicates a dissolution of Al from the crucible and a formation of Magnesia-Alumina-Spinel.

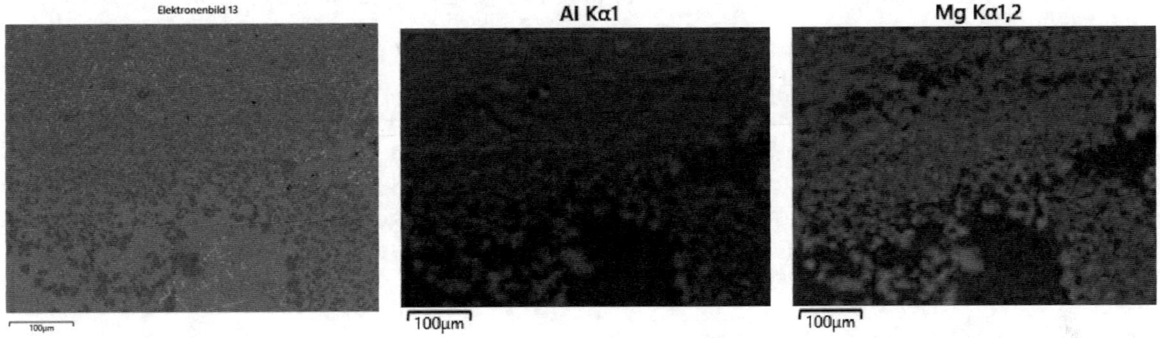

Figure 6. SEM image, Mg and Al EDX-distribution of EAF-sat slag at the crucible surface-near region

It must be mentioned that in this publication, only selected results are shown to keep the scope within limits. Nevertheless, it should be enough to demonstrate the method and compare the different slags roughly.

4. DISCUSSION

The EAF and EAF-sat slags show outstanding foaming behavior. Although slight differences in their morphology can be seen, no significant difference in their forming height is observed. That fact does not agree with the literature [4,16,17], where a low but limited number of solid particles is described as foaming enhancing. Since we had no access to XRD analysis, we cannot exclude precipitation effects during the foaming tests that influence the behavior. However, as can be seen in figure 7, the almost horizontal phase border between $Slag_{liq}$ and $Slag_{liq}$ + (Mg,Fe)O indicates this influence to be small for short forming times. The Smelter-like slags do not show any foamability, even with increased (FeO) content, which results in a higher amount of {CO}. A further increase in (FeO) could be tested. However, as iron oxide reduces slag viscosity, one can expect a similar result.

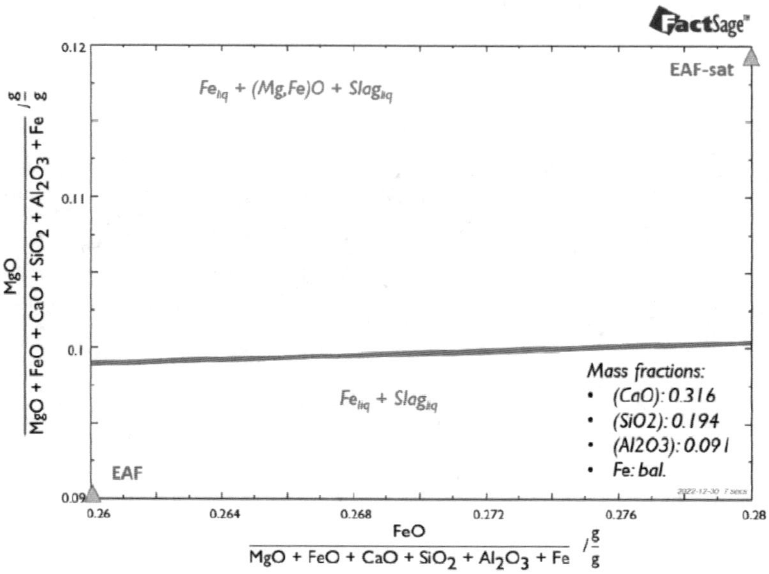

Figure 7. SEM image, Mg and Al EDX-distribution of EAF-sat slag at the crucible surface-near region [18]

From a method perspective, all tests are considered successful. Two weaknesses can be seen in the limited thermal conductivity and the interactions with the refractory material. However, the latter may explain the poor correlation between laboratory results and industrial conditions, e.g., noticed by de Vos et al. [19]. One possibility to eliminate these weaknesses would be to use metallic crucibles in an inert atmosphere. Although a quantitative transformation to large-scale EAF may be limited, the method should be applicable for qualitative analyses of, e.g., different slag compositions or carbon carriers.

5. CONCLUSIONS AND OUTLOOK

The developed method enables testing the slag foaming behavior with minimal effort on a laboratory scale. Although some weaknesses, such as interactions with the crucible or the limited thermal conductivity of the refractory, reduce the significance, it can ideally be applied for comparing different specifications like different carbon sources or slag mixtures with each other. From a slag perspective, the tested EAF slags show outstanding foamability, whether saturated or not. However, Smelter-like ones do not foam in the tested FeO-range, of about 5-10 %.

References

[1] Abel, M. Elektrische Auslegung von AC Lichtbogenöfen. Online Seminar, *VDEH – Elektrotechnik des Lichtbogens,* March 16, 2021.

[2] Zhan, D.; Wang, J.; Huang, L.; Zhang, H. Effect of Slag Compositions on Change Behavior of Nitrogen in Molten Steel. *Metals* **2022**, *12*, 846, doi:10.3390/met12050846.

[3] The Making, Shaping and Treating of Steel: Steelmaking and Refining Volume; Fruehan, R.J., Ed., 11th ed., 1998.

[4] Pretorius, E.B.; Carlisle, R.C. Foamy slag fundamentals and their practical application to electric furnace steelmaking. *Iron and Steelmaker(USA)* **1999**, *26*, 79–88.

[5] Ito, K.; Fruehan, R.J. Study on the foaming of CaO-SiO2-FeO slags: Part I. Foaming parameters and experimental results. *Metall and Materi Trans B* **1989**, *20*, 509–514,

doi:10.1007/BF02654600.

[6] Ito, K.; Fruehan, R.J. Study on the foaming of CaO-SiO2-FeO slags: Part II. Dimensional analysis and foaming in iron and steelmaking processes. *Metall and Materi Trans B* **1989**, *20*, 515–521, doi:10.1007/BF02654601.

[7] Zhang, Y.; Fruehan, R.J. Effect of the bubble size and chemical reactions on slag foaming. *Metall and Materi Trans B* **1995**, *26*, 803–812, doi:10.1007/BF02651727.

[8] Ozturk, B.; Fruehan, R.J. Effect of temperature on slag foaming. *Metall and Materi Trans B* **1995**, *26*, 1086–1088.

[9] Zhang, Y.; Fruehan, R.J. Effect of carbonaceous particles on slag foaming. *Metall and Materi Trans B* **1995**, *26*, 813–819, doi:10.1007/BF02651728.

[10] Jung, S.-M.; Fruehan, R.J. Foaming Characteristics of BOF Slags. *ISIJ International* **2000**, *40*, 348–355, doi:10.2355/isijinternational.40.348.

[11] Jiang, R.; Fruehan, R.J. Slag foaming in bath smelting. *Metall and Materi Trans B* **1991**, *22*, 481–489, doi:10.1007/BF02654286.

[12] Ghag, S.S.; Hayes, P.C.; Lee, H.-G. The Prediction of Gas Residence Times in Foaming CaO-SiO2-FeO Slags. *ISIJ International* **1998**, *38*, 1216–1224, doi:10.2355/isijinternational.38.1216.

[13] Ghag, S.S.; Hayes, P.C.; Lee, H.-G. Physical Model Studies on Slag Foaming. *ISIJ International* **1998**, *38*, 1201–1207, doi:10.2355/isijinternational.38.1201.

[14] Ghag, S.S.; Hayes, P.C.; Lee, H.-G. Model Development of Slag Foaming. *ISIJ International* **1998**, *38*, 1208–1215, doi:10.2355/isijinternational.38.1208.

[15] Echterhof, T.; Pfeifer, H. Potential of biomass usage in electric steelmaking, 2011. *1st EECR Steel conference*

[16] Heo, J.H.; Park, J.H. Assessment of Physicochemical Properties of Electrical Arc Furnace Slag and Their Effects on Foamability. *Metall and Materi Trans B* **2019**, *50*, 2959–2968, doi:10.1007/s11663-019-01671-4.

[17] Kim, H.S.; Min, D.J.; Park, J.H. Foaming Behavior of CaO-SiO2-FeO-MgOsatd-X (X=Al2O3,MnO, P2O5, and CaF2) Slags at High Temperatures. *ISIJ International* **2001**, *41*, 317–324, doi:10.2355/isijinternational.41.317.

[18] Pfeiffer, A. Evaluation of the Smelting Behavior of Direct Reduced Iron. Phd thesis, Montanuniversität Leoben, 2023

[19] Vos, L. de; Cnockaert, V.; Bellemans, I.; Vercruyssen, C.; Verbeken, K. Critical Assessment of the Applicability of the Foaming Index to the Industrial Basic Oxygen Steelmaking Process. *steel research int.* **2021**, *92*, 2000282, doi:10.1002/srin.202000282.

IOP Conf. Series: Materials Science and Engineering 1309 (2024) 012008 doi:10.1088/1757-899X/1309/1/012008

Manipulation of the leaching behaviour of high-alloyed steel EAF slags through the particle size modulation

D Mombelli[1], G Dall'Osto[1], S Cozzo[1], C Mapelli[1]

[1] Politecnico di Milano, Dipartimento di Meccanica, Via La Masa 1, 20156 Milano, Italy

davide.mombelli@polimi.it

Abstract. The safe disposal of EAF slag have become crucial due the environmental and health protection regulations. Its direct reuse is hampered by the possible release of toxic elements into the surrounding environment over time. Specifically, EAF slag from high-alloyed steel production could potentially release toxic metals (e.g., Ba, Cr, Mo and V). Currently, the assessment of leaching behaviour is carried out according to the EN 12457:2002 standard, which generically specify that the material should not be finely ground, limiting the particle size as much as possible. However, no guidance is given on reducing particle size and handling fine fractions, which could lead to manipulation of data and variability of results. In this work six different fractions (from <4 mm to <63 μm) of EAF slag from high-alloyed steel production were investigated. X-Ray Diffraction, Scanning Electron Microscopy and selective C_2S dissolution were used to correlate the leaching behaviour with particle size and identify which mineral can alters the release of specific toxic elements. The pH of the water was measured immediately after filtration, and the solution was analysed by Inductively Coupled Plasma-Optical Emission Spectrometry. In addition, five mixtures were prepared using different mass fractions of the six granular fractions with the aim of unbalancing the mixtures toward coarser or finer fractions, respectively. All the mixtures were subjected to leaching under the same previous conditions and analysed to evaluate the behaviour of the different blended fractions. To investigate the predictability of the leaching behaviour of the mixtures, the experimental results were compared with the values predicted by a theoretical model calibrated using the results obtained from the leaching results of the six particle size fractions. The investigation showed that the leaching behaviour is negatively affected by the particle size reduction, mainly due to the increase of surface-to-volume ratio, which increases the dissolution rate. Similarly, the mass fraction of the main crystalline phases that regulate leaching (unstable spinels, brownmillerite and larnite) is also correlated with the change in particle size. Consequently, the leaching results for the mixtures highlighted a critical aspect contained in the European standard. Even small changes in particle size can lead to different concentrations of toxic metals in leachates, which can then be manipulated to achieve the desired result.

1. Introduction

In recent years, the safe disposal and monitoring of steel slag has become a key issue as a consequence of the environmental and health protection regulations. Electric Arc Furnace (EAF) slag could be exploited in several application fields, such as road construction and concrete production. However, its use can be hampered by the presence and possible release of toxic elements into the surrounding environment [1]. In particular, EAF slag from high-alloyed steel production could potentially release toxic metals such as Barium (Ba), Chromium (Cr), Molybdenum (Mo) and Vanadium (V).

Currently, the assessment of leaching behaviour is carried out through specific test, performed according to the EN 12457:2002 standard. In detail, according to the guidelines of the European standards, it is commonly specified that the material should not be finely ground, limiting the particle size as much as possible, without giving any indication of particle size reduction and the handling of 'fine fractions'. However, as leaching behaviour is influenced by various chemical and physical parameters such as pH, temperature, liquid-solid ratio and particle size [2–6], hence the possibility of data manipulation and variability of results cannot be excluded and relies on the operator. Consequently, the choice of particle size is crucial as it can heavily influence the results of the leaching test. This issue has already been raised in 2013 by Volante et al. [7] and partially resolved through the emanation of a specific protocol, valid only in Italy, that specifies the methodology to follow for the sample preparation prior to conducting the leaching test according to EN 12457:2002 [8]. However, this protocol consider as "fine" all the fraction passing through the 0.5 mm sieve and does not provide any quantitative information about the "weight" of the different granularity fraction on the final leachate toxic elements concentrations.

For this reason, in this work, six different granularity fractions of EAF slag from high-alloyed steel production were investigated to correlate the leaching behaviour with the particle size. Mineralogical, chemical and microstructural characterisations were performed to identify the main structural constituents of the slag. Characterisation results were correlated with the results of standard leaching tests (in accordance with EN 12457-2:2002). In addition, standard leaching tests were also performed on five granulometric mixtures to assess the effect of particle size distribution on the leaching behaviour of a natural-like granular material.

The main aim of this work is to identify which mineral phase is responsible for the toxic elements leaching and to assess the effect of particle size reduction on the leaching behaviour. This survey allowed to identify successfully how the particle size reduction influences both the nature and the distribution of the main crystalline phases which, in turn, directly influence the leaching of pollutants.

2. Materials and methods

The EAF slag was supplied by an Italian electric steelwork and is associated with high-alloyed steel production. The average chemical composition of the investigated slag is reported in **Table 1**.

Table 1. Average chemical composition of investigated EAF slag (wt.%).

MgO	Al_2O_3	SiO_2	CaO	FeO	Cr_2O_3	MnO
8.22	3.64	9.80	30.62	29.76	12.00	5.46

The investigated EAF steel slag was initially sieved and partitioned into six different granularity fractions (**Table 2**). Each fraction was characterised mineralogically, chemically and microstructurally by means of X-Ray Diffraction (XRD) and Scanning Electron Microscopy (SEM) analyses. Standard leaching tests were performed according to EN 12457-2 on each granularity fraction in deionised water at pH 7. All the samples with a particle size greater than 4 mm were not considered as the standard requires a particle size below this value. Leaching test results were correlated to the characterisation previously performed.

Table 2. Sample classification according to the particle size.

Sample	B	C	D	E	F	G
Particle size (mm)	2-4	1-2	0.5-1	0.125-0.5	0.063-0.125	<0.063

X-ray diffraction experiments were performed with a Rigaku SmartLab SE multipurpose X-ray diffractometer in a θ-θ configuration, operating at 40 kV and 40 mA. The samples were scanned between 5° and 90° with a scan rate of 1°/min. Both qualitative and quantitative analyses were performed with SmartLab Studio II software by means of Crystallographic Open Database (COD). To perform the quantitative analysis, the Rietveld refining method was used [9].

To verify some of the results obtained through the Rietveld refinement, C_2S selective dissolution was experimentally performed on all samples. 5 g of slag powder was mixed in 300 ml of methanol with 20 g of salicylic acid. The solution was stirred for 2 hours and then filtered in a 0.45 μm vacuum system. The residual slag powder was dried and weighed again to estimate the larnite fraction [9]. Chemical and microstructural characterisation was performed by Zeiss Sigma 500 Field Emission Scanning Electron Microscopy (FE-SEM) equipped with the Oxford Instruments Ultim Max 65 detector for Energy Dispersive X-ray (EDX) analysis. The leaching behaviour of each granularity fraction was assessed through standard leaching tests, carried out according to EN 12457-2. The test consists in submerging 45 g of slag in a fixed volume of deionised water (10 l/kg) and stirring the solution at 10 rpm for 1 day. After 24 hours, slag is removed, the solution vacuum filtered using a 0.45 μm filter and water pH measured instantly after. The solution is then analysed by Inductively Coupled Plasma - Optical Emission Spectrometry (ICP-OES) focusing on the elemental concentration of Ba, Cr, Mo, V and W.

In addition, five granulometric mixtures were prepared by using different mass fractions of the six samples (**Table 3**). The mass fractions were chosen to unbalance the mixtures towards coarser or finer fractions. All mixtures were leached in the same conditions as before and analysed to assess the leaching behaviour of different fractions mixed and to find out whether the leaching behaviour can be predicted.

Table 3. Mass composition (wt.%) of the five granulometric mixtures investigated.

	%B	%C	%D	%E	%F	%G
MIX 1	35	35	10	10	5	5
MIX 2	5	5	10	10	35	35
MIX 3	10	10	35	35	5	5
MIX 4	5	5	35	35	10	10
MIX 5	20	20	20	20	0	20

3. Results

The crystallographic analysis highlighted the occurrence of three main phases in all the samples: chromite (belonging to the spinel family, mainly as $(Fe,Mn,Mg){\cdot}(Al,Cr)_2O_3$ spinel), brownmillerite ($4CaO{\cdot}Al_2O_3{\cdot}Fe_2O_3$) and larnite ($2CaO{\cdot}SiO_2$). The presence of scheelite (Ca_2MgO_6W) was detected in all samples as minor phase. Furthermore, in samples C, D, E and F, calcium-chromium oxide phase ($CaCr_2O_4$) was also found (detected as a slight peak at $2\theta = 16°$). **Table 4** shows the mass fractions of the three main crystalline phases, obtained through the Rietveld refinement method. In accordance with Herbelin et al. [10], brownmillerite abundance decreases with the particle size reduction, whereas spinel solid solution (chromite) follows the opposite trend. Brownmillerite, being a resistant phase, is preferentially found in coarse grains and probably reflects the initial composition of the slag. Contrary to what was found by Herbelin et al. [10], larnite abundance is almost constant for each sample. The larnite fraction measured experimentally by the selective dissolution technique is also reported (L_{diss}) in **Table 4**. Since the results of selective dissolution are consistent with the theoretical larnite values, the XRD quantitative analysis was considered reliable. Only sample F showed a different trend which can be attributed to its preparation procedure. Indeed, during the particle size reduction process, sample F was obtained with some difficulty as the ground fraction remained of a slightly coarser particle size (sample E) or passed through the sieve to form the finest fraction (sample G). To obtain sample F, therefore, several cycles of ball milling were performed on the slag. Consequently, sample F should be considered as an "artificial granularity fraction" rather than a "natural granularity fraction". It is likely that this granularity fraction consists of small pieces from the filing of coarser fractions, which could not be further reduced in size having probably already reached their size limit.

Table 4. XRD quantitative analysis by Rietveld method and larnite selective dissolution results (wt.%).

	Chromite	Brownmillerite	Larnite	L_{diss}
B	43.07	31.79	21.15	21.00
C	46.46	27.90	21.26	21.20
D	48.09	25.64	21.53	21.50
E	49.00	23.00	22.25	22.20
F	45.00	33.30	18.80	18.80
G	49.90	21.90	22.80	22.80

By SEM-EDS analysis, the mapping of toxic metals was performed on all samples. Ba and Mo were not localised, probably because their concentrations were below the detection limit of the EDS probe. Chromium is mainly dissolved in spinels (56-61 at.%) and is also associated, to a lesser extent, with brownmillerite (max 10 at.%) and scheelite (max 8 at.%). Traces of Cr (less than 0.5 at.%) are also found in larnite (only sample F showed a higher Cr concentration, equal to 2.5 at.%) [11,12]. The concentration of chromium in both spinels and brownmillerite tends to increase by decreasing the particle size, except in sample F. Vanadium is dispersed almost everywhere. The highest concentration of V is always found in brownmillerite (max 2 at.%) [11,12]. Coarser granularity fractions showed higher V concentration in this phase compared to smaller samples (E and G). Sample F showed the highest V concentration in brownmillerite. Tungsten is mainly dissolved in scheelite; traces of W are also found in larnite and brownmillerite. A higher concentration of tungsten in both scheelite and larnite was found for smaller granularity fractions. Finally, the concentration of calcium in both larnite and spinels increases as the particle size decreases.

Table 5 shows the results of standard leaching tests performed on all six granularity fractions. The data are reported in ppm (mg/l). To better visualise the effect of particle size, **Figure 1** shows the leaching results as a function of the average size of each sample ($p/A_{(50)}$).

Table 5. Results of standard leaching test performed on each sample (mg/l).

	Ba	Cr	Mo	V	W	pH
B	0.13	2.20	0.18	1.74	0.93	11.33
C	0.15	2.35	0.22	1.76	0.93	11.41
D	0.18	3.14	0.25	2.00	1.08	11.57
E	0.26	3.58	0.36	1.18	1.10	11.95
F	0.23	3.60	0.27	2.65	1.64	11.80
G	0.29	5.17	0.48	0.69	1.80	12.15

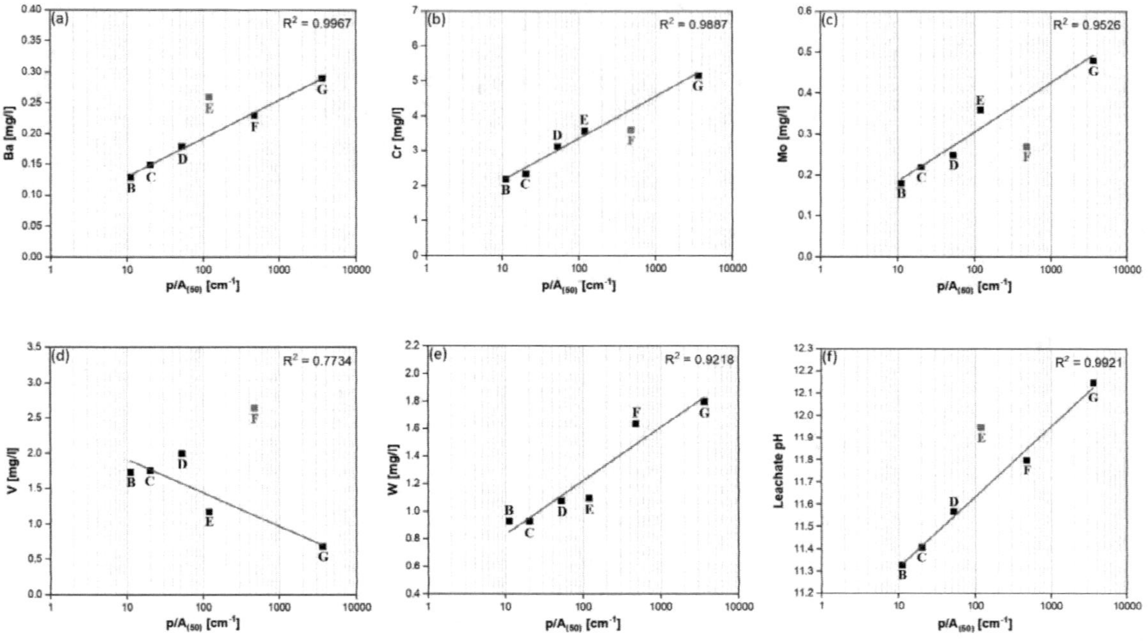

Figure 1. Standard leaching test results Vs p/$A_{(50)}$.

For all elements (except V) and pH, the logarithmic regression shows that the general trend is good, with a coefficient of determination (R^2) always above 92%. The regression was performed in most cases on all samples, except for some outliers (samples E or F, highlighted in red compared to the others).

The general trend of the results shows that leaching increases as the particle size decreases, except in the case of V. In general, the highest concentrations are found in the finer granularity fractions. This is mainly due to their higher surface-to-volume ratio: by increasing this latter, the dissolution rate also increases.

Correlation between leaching test results and mineralogical and chemical characterisation can allow to identify some aspects ruling the interaction between the slag and the water and to understand how the particle size directly affects it.

3.1. Cr leaching

Concerning Cr leaching, spinel formation is usually a good way to fix chromium and prevent its leaching [13]. However, in high basicity slag, soluble Cr-bearing phases can be formed, such as unstable spinels. In this condition, Cr can be leached easily, even if it is bound in a spinel structure [12]. The simplest way to assess Cr leaching is to focus on slag basicity (in terms of CaO/SiO_2 ratio). For the investigated slag, the ratio is equal to 3.1 and therefore the slag can be considered strongly basic, which leads to unstable spinels and increased Cr leaching [12].

According to **Table 5**, Cr leaching increases as the particle size decreases. This is mainly due to the grinding procedure, which causes a higher mass fraction of chromite by reducing the particle size (**Figure 2a**). Furthermore, also the spinel size strongly influences Cr leaching, since their dimensional reduction is proportional to the slag granulometry. In other words, for smaller samples are more prone to be oversaturated in Cr and, thus, highly soluble. **Figure 2b** shows Cr leaching as a function of the average area of spinels. The reduction in spinel size explains why sample F showed higher Cr leaching compared to coarser granularity fractions, even though the chromite mass fraction is lower for this sample. In addition, the higher Cr concentration in larnite could have increased its leaching for sample F.

Finally, the effect of pH must also be considered. Chromium concentration is typically manifested only in caustic environments [2,3]. The particle size reduction causes an increase in the leachate pH, leading to a higher Cr leaching (**Figure 2c**).

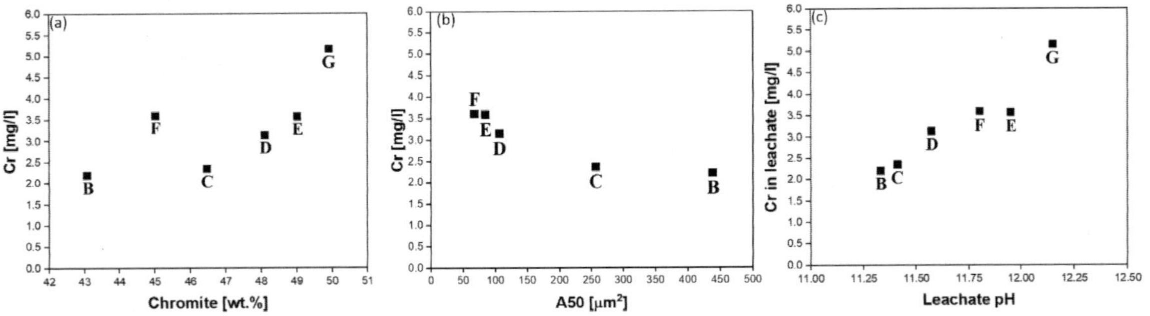

Figure 2. Cr leaching vs (a) chromite (wt.%); (b) spinel size and (c) leachate pH.

3.2. V leaching

According to **Table 5**, V leaching increases up to sample D. For smaller samples, a decrease in its leaching is observed, except for sample F which showed the highest V leaching. Particle size reduction influences V leaching by changing the mass fraction of brownmillerite. Coarser granularity fractions, characterised by higher mass fractions of brownmillerite, showed a high V leaching (**Figure 3a**) compared to finer granularity fractions (such as samples E and G). Only sample F seems to be outside this trend, showing the highest V leaching, mainly due to the highest mass fraction of brownmillerite and the highest concentration of vanadium within.

The leaching of vanadium is also influenced by larnite. Fine granularity fractions, characterised by higher mass fractions of larnite, showed less V leaching compared to coarser granularity fractions (**Figure 3b**). Larnite is a highly soluble phase and its dissolution is a key aspect as it may determine the final pH of the leachate. The dissolution of a higher amount of larnite in the finer fractions can bring more Ca^{2+} and OH^- into the solution, causing a strong increase in the pH compared to coarser fractions. **Figure 3c** shows how V leaching intimately responds to the pH changes, being considerably reduced when pH rises towards 12. Vanadium oxide is considered an amphoteric species and, in alkaline solution, tends to behave as acid oxide. This explains the high concentration of V in alkaline leachates (pH in the range of 11–11.80). However, when pH reaches very high values, the solubility of vanadium is drastically reduced [4].

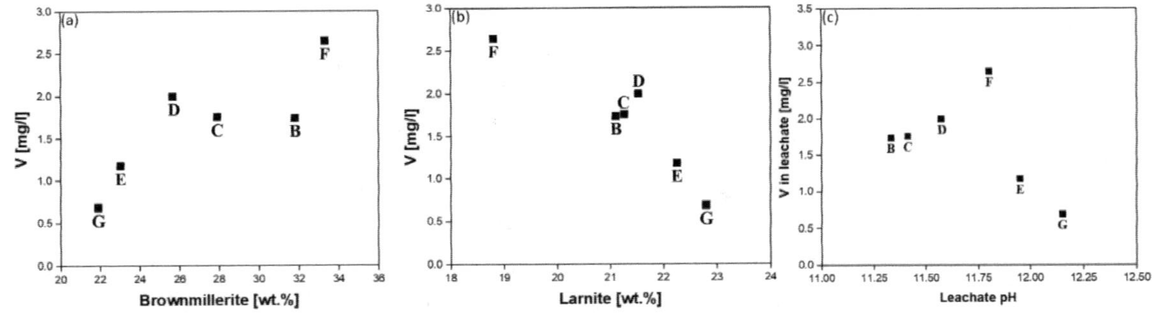

Figure 3. V leaching vs (a) brownmillerite (wt.%); (b) larnite (wt.%) and (c) leachate pH.

3.3. Ba leaching

The literature suggests that barium is typically combined with calcium silicates. As stated by Mombelli et al. [4], if the slag contains enough Mg-complexing oxides (such as FeO), barium can dissolve in larnite. Indeed, as highlighted by **Figure 4a** Ba leching is mainly controlled by larnite: in general, the higher the mass fraction of larnite, the higher the leaching of barium. As in the case of spinels, the phase size of larnite must also be considered. **Figure 4b** shows Ba leaching as a function of the average area of larnite grains. As in the previous observations, the larnite grains dimensions follows the slag granulometry and thus more soluble in the smaller fractions. A logarithmic trend was identified between the larnite fraction and the final water pH (**Figure 4c**). This relation confirms the key role played by larnite in controlling the leaching behaviour: since larnite determines most of the Ba and Ca concentration in the leachate, it also determines the final pH of the solution [11].

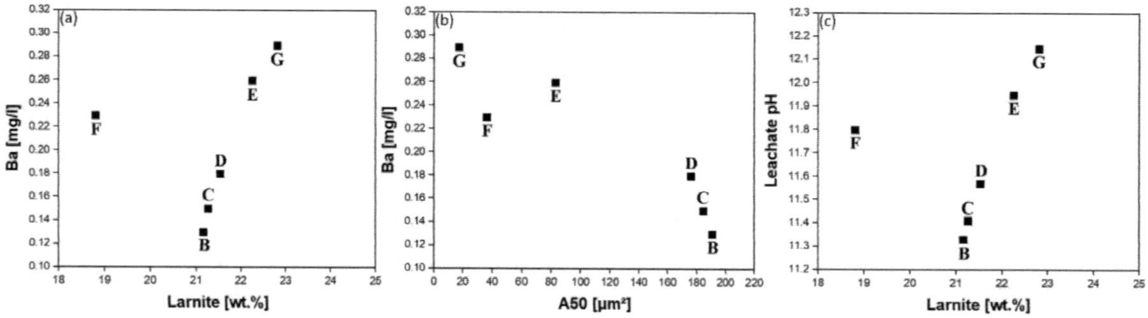

Figure 4. Ba leaching vs (a) larnite (wt.%) and (b) larnite size. Leachate pH vs larnite (wt.%) in (c).

3.4. W and Mo leaching

Concerning Mo leaching, the literature suggests that Mo is dissolved in iron-containing phases such as brownmillerite [14,15]. The higher leaching of Mo in the smaller samples can be attributed to the larger surface area of the finer granularity fractions which imposes a higher reactivity, even though the mass fraction of brownmillerite is lower. In addition, pH also influences Mo leaching: the higher the pH, the higher the concentration of Mo in the leachate (**Figure 5a**).

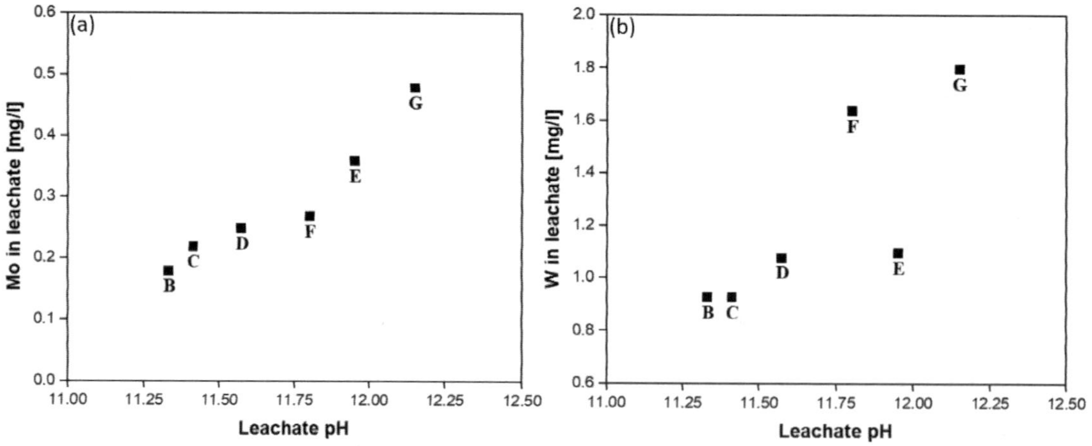

Figure 5. Mo (a) and W (b) leaching vs pH.

No extensive information are present in literature regarding W leaching. Since scheelite contains most of the tungsten, it is likely to control its leaching. It is reasonable to assume that the size of the scheelite decreases with the particle size. This could lead to a higher W leaching in the smaller granularity fractions, mainly due to the higher surface-to-volume ratio which should increase the dissolution rate. In addition, the concentration of tungsten in both larnite and scheelite increases in the smaller granularity fractions. This could further explain why smaller samples showed higher W leaching compared to coarser ones. As stated by Quadrio et al.[16], the leaching of tungsten tends to grow linearly with the CaO content in the slag. Although, in this work, the general chemical composition of each granularity fraction was not calculated, it is possible to state that higher CaO content should be detected in smaller granularity fractions, when compared to coarser ones, due to the higher amount of Ca contained in both spinels and larnite. Therefore, the higher W leaching for the smaller granularity fractions could have been increased by their higher CaO content which leads to higher pH values. This is actually confirmed by the general increase in W leaching by increasing the leachate pH (**Figure 5b**).

3.5. *Leaching test on mixtures*

Table 6 shows the results of standard leaching tests performed on the five granulometric mixtures. The data are reported in ppm (mg/l). The results of the mixtures are consistent with the trend of the results of the granularity fractions (**Figure 6**).

On the one hand, mixtures 1 and 2 can be labelled as the extremes of the leaching behaviour, the first showed the lowest leaching due to the lowest pH, the second the highest leaching due to negative influence of the finest granularity fractions. On the other hand, mix 5, due to its balance on almost all fractions, showed a intermediate leaching behaviour. Finally, mixtures 3 and 4 agree for the pH but showed slightly different leaching compared to mix 5, probably due to the natural variability that the leaching test inherently possesses. Leaching results demonstrate the critical issue contained in the European Standard. Even small changes in the particle size can lead to different toxic metal concentrations in the leachates.

Table 6. Results of leaching tests on mixtures (mg/l).

	Ba	Cr	Mo	V	W	pH
Mix 1	0.18	2.47	0.24	2.03	1.21	11.61
Mix 2	0.24	4.00	0.33	1.51	1.57	11.95
Mix 3	0.19	2.70	0.27	1.92	1.03	11.75
Mix 4	0.20	2.85	0.28	1.93	1.09	11.80
Mix 5	0.20	3.05	0.32	1.84	1.27	11.72

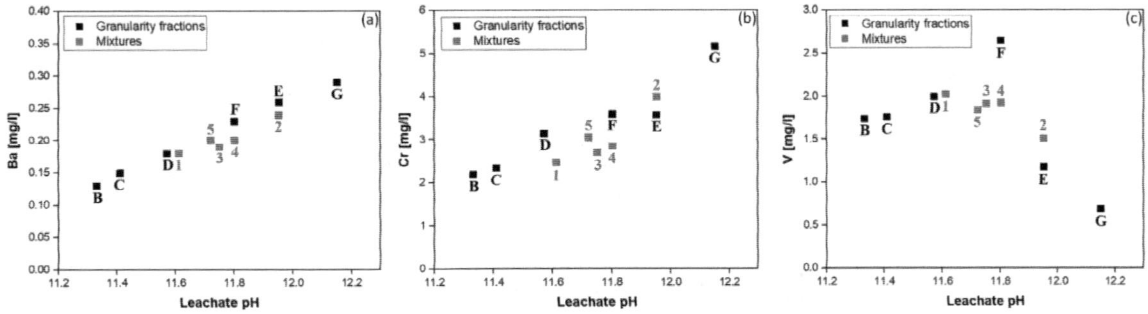

Figure 6. Leaching of (a) Ba; (b) Cr and (c) V vs leachate pH for mixtures.

4. Discussion

The results showed that, in general, finer granularity fractions negatively influence the leaching behaviour, leading to a higher leaching of elements. This is further confirmed by the results of mixture 2 (the mixture with the majority of fine fractions, namely F and G), which showed higher elemental leaching than the other mixtures, except for vanadium.

As a consequence, it is theoretically possible to create and use suitable particle size mixtures according to the operational needs (e.g., if required in road construction), but in general these mixtures should still contain larger particle size fractions, as they show better leaching behaviour. Nevertheless, the investigated EAF slag cannot be employed even if 100% of the largest granularity fraction (sample B, 2-4 mm) is used. In fact, according to the results of standard leaching test (**Table 5**), this fraction showed a concentration of elements in leachate well above the imposed national limits. Therefore, to be able to use this slag, some stabilisation treatments, aimed to modify the chemical composition of the slag, are required to transform it into a safe and environmentally friendly artificial aggregate [17].

The leaching tests performed on the granulometric mixtures demonstrated the critical issues contained in the currently used European Standard (EN 12457:2002), which does not provide any indication about particle size reduction and management of "fine fractions". Indeed, particle size reduction strongly influences the leaching behaviour, especially by changing the pH of the final leachate. This is well demonstrated by the five mixtures for which even small changes in the mass of the granularity fractions lead to different toxic metal concentrations in the leachate. Therefore, particle size and the management of fine fractions are key aspects to consider when assessing slag leaching behaviour.

5. Conclusions

In this work, the effect of particle size on the leaching behaviour of EAF slag from high alloyed steel production was investigated. The conclusions of this study can be summarized as follows:

- XRD and SEM analyses showed that unstable spinels, brownmillerite and larnite are the main crystalline phases ruling the leaching of toxic metals.
- In general, finer fractions negatively influence the leaching behaviour, mainly due to the higher surface-to-volume ratio which increases the dissolution rate.
- The particle size reduction strongly influences the distribution (in terms of size and mass fraction) of the crystalline phases.
- Unstable spinels mainly rule chromium dissolution. The higher mass fraction of spinels in the smaller granularity fractions leads to higher Cr leaching. In addition, smaller spinels increase Cr leaching.
- The leaching of V is mainly controlled by brownmillerite. Coarser granularity fractions showed higher V leaching compared to finer granularity fractions. The 0.063 – 0.125 mm fraction (sample F) showed the highest V leaching.
- Larnite is responsible for Ba leaching. The higher mass fraction of larnite in the smaller granularity fractions leads to higher Ba leaching. In addition, smaller grains of larnite increase Ba leaching.
- Larnite influences V leaching, determining the pH of the final leachate. Very high pH values drastically reduce V leaching.
- Leaching tests performed on mixtures showed that the leaching behaviour can be predicted. Leaching results demonstrated the critical issues contained in the European Standard.
- The main effect of particle size reduction is to change the final leachate pH which, in turn, strongly influences the leaching of metals under investigation.

References

[1] Fisher L V. and Barron A R 2019 The recycling and reuse of steelmaking slags — A review *Resour. Conserv. Recycl.* **146** 244–55
[2] van der Sloot H A and Mulder E 2002 *Test Methods To Assess Environmental Properties of Aggregates in Different Applications : the Role of En 1744-3*
[3] van der Sloot H A and Dijkstra J J 2004 *Development of horizontally standardized leaching tests for construction materials: a material based or release based approach?* (Petten)
[4] Mombelli D, Mapelli C, Barella S, Di Cecca C, Le Saout G and Garcia-Diaz E 2016 The effect of chemical composition on the leaching behaviour of electric arc furnace (EAF) carbon steel slag during a standard leaching test *J. Environ. Chem. Eng.* **4** 1050–60
[5] Quina M J, Bordado J C M and Quinta-Ferreira R M 2009 The influence of pH on the leaching behaviour of inorganic components from municipal solid waste APC residues *Waste Manag.*

29 2483–93
[6] Komonweeraket K, Cetin B, Aydilek A H, Benson C H and Edil T B 2015 Effects of pH on the leaching mechanisms of elements from fly ash mixed soils *Fuel* **140** 788–802
[7] Volante M, Marmai S, Attuati S, Rinaldi L, Guarneri R, Amico F and Calvisi F 2013 La macinazione sotto 4 mm dei materiali granulari per il test di cessione secondo le norme UNI EN 12457-2 e UNI 10802. Una questione ancora aperta. *BEA Boll. degli esperti Ambient.* **4**
[8] Schweiger A, Balzamo S, Bettinelli M, Corbella G, Corsini A, De Zorzi P, Massimiliano A, Pedersini P, Perego G, Porro E M A and Vizzini G 2020 UNI/PdR 94:2020 - Scoria nera da forno ad arco elettrico (EAF) - Metodo per la preparazione del campione da sottoporre a prova di lisciviazione secondo la UNI EN 12457-2 1–24
[9] Stutzman P E, Feng P and Bullard J W 2016 Phase analysis of portland cement by combined quantitative X-ray powder diffraction and scanning electron microscopy *J. Res. Natl. Inst. Stand. Technol.* **121** 47–107
[10] Herbelin M, Bascou J, Lavastre V, Guillaume D, Benbakkar M, Peuble S and Baron J P 2020 Steel slag characterisation—benefit of coupling chemical, mineralogical and magnetic techniques *Minerals* **10** 1–19
[11] Mombelli D, Mapelli C, Barella S, Di Cecca C, Le Saout G and Garcia-Diaz E 2016 The effect of microstructure on the leaching behaviour of electric arc furnace (EAF) carbon steel slag *Process Saf. Environ. Prot.* **102** 810–21
[12] Mombelli D, Barella S, Gruttadauria A, Mapelli C, Le Saout G and Garcia-Diaz E 2018 Effects of basicity and mesh on Cr leaching of EAF carbon steel slag *Appl. Sci.* **9**
[13] Kühn M, Mudersbach D, Baena Liberato J M, De Angelis V, Capodilupo D and de Fries U 2006 Chrome immobilisation in EAF slags from high-alloy steelmaking : development of a slag treatment process *Rep. Comm. Eur. Communities - EUR, Tech. steel Res.* **22077** 1–96
[14] Kaja A M, Delsing A, van der Laan S R, Brouwers H J H and Yu Q 2021 Effects of carbonation on the retention of heavy metals in chemically activated BOF slag pastes *Cem. Concr. Res.* **148**
[15] Spanka M, Mansfeldt T and Bialucha R 2018 Sequential extraction of chromium, molybdenum, and vanadium in basic oxygen furnace slags *Environ. Sci. Pollut. Res.* **25** 23082–90
[16] Mombelli D, Quadrio M, Mapelli C, Echterhof T, Aparicio L F, Sorger C, Griessacher T, Mudersbach D and Schüler S 2022 Influence of the Recirculation of Various by-products Generated through Electric Arc Furnace Route on EAF Slag Quality *ISIJ Int.* **62** 2610–21
[17] Mombelli D, Mapelli C, Barella S, Gruttadauria A, Saout G Le and Garcia-diaz E 2014 The efficiency of quartz addition on electric arc furnace (EAF) carbon steel slag stability *J. Hazard. Mater.* **279** 586–96

IOP Conf. Series: Materials Science and Engineering 1309 (2024) 012009 doi:10.1088/1757-899X/1309/1/012009

Decreasing the environmental impact of the electric steel making process through implementation of furnace retrofitting solutions

F. Kaiser[a], T. Reichel[a], T. Echterhof[a], D. Mier[b], I. Unamuno[c], H. Pfeifer[a]
a) RWTH Aachen Department for Industrial Furnaces and Heat Engineering
b) Sidenor Investigación y Desarrollo S.A.
c) Sidenor Aceros Especiales S.L

kaiser@iob.rwth-aachen.de

Abstract. Steel production in the electric arc furnace is an energy-intensive process and generates a high amount of climate-relevant CO_2 emissions. During the last decades, various improvements of the furnace and its operation have been developed and implemented, so that the efficiency of modern steel plants has reached a high level. Due to the standardized process flows and high production volumes, even small improvements in resource efficiency can lead to significant savings and improve the environmental impact of the steel production. This paper presents several optimization activities and retrofitting solutions for the metallurgical sector and in particular for the electric steel route. On the example of a Spanish electric steel plant, a life cycle analysis is carried out to calculate the potential improvements in environmental impacts and compare them with the current state of the art. For this purpose, real process data were collected and transferred to a life cycle assessment model with the use of a material flow analysis. A comparison of the ecological impact categories provides information on which retrofitting solutions can improve the process in the electric arc furnace and ensure a lower environmental impact.

1. Introduction

The presented work was carried out over the course of the project "Retrofitting Equipment for Efficient Use of Variable Feedstock in Metal Making Processes" (REVaMP), funded under the European Union's Horizon 2020 research and innovation program. This paper focuses specifically on the work carried out until now for the use case of an electric arc furnace (EAF) plant for steel scrap recycling. An explanation of the overall objectives and actions of this project are given as well.

The main objective of the REVaMP project is to develop, adapt and apply novel retrofitting technologies to cope with the increasing variability and to ensure an efficient use of the feedstock in terms of materials and energy. The developed retrofitting technologies are tested under usual production circumstances. At the end of the project the performance of each retrofitted use case is evaluated regarding technical, economic and environmental criteria. The analysis presented in this article focuses on the environmental evaluation of the selected steel use case in particular, and describes the current state of the process (base line) and aspects of its current ecological evaluation as well as the potential impact that the retrofitting solutions might have on this evaluation.

IOP Conf. Series: Materials Science and Engineering 1309 (2024) 012009 doi:10.1088/1757-899X/1309/1/012009

2. Materials and methods

First, the technologies developed over the course of the REVaMP project are explained in more detail. To better adjust the processes included in the project to a variable feedstock, the production plants of the industrial partners shall be retrofitted with appropriate sensors for scrap analysis and furnace operation. Furthermore, the selection of the optimal feedstock in terms of material and energy efficiency shall be improved by application of appropriate process control and decision support tools. To monitor and control the process behaviour in an optimal way, model-based software tools are developed and applied. [3]

The retrofitting solutions are exemplarily demonstrated within three different use cases from the metal making industry, namely electric and oxygen steelmaking, aluminium refining and lead recycling. The performance of the different technologies are assessed, and the benefits are evaluated in terms of economic and ecological effects, as well as cross-sectorial applicability in other process industries.

The retrofitting technologies can be divided into four categories: new sensors, new equipment for efficient energy usage, new models for feedstock tracking and new tools for process monitoring, control and optimisation. Due to the multitude of technologies, only the ones planned for the electric arc furnace process are now explained in more detail. The measures planned for the EAF plant consist of a scrap mix and energy efficiency optimisation systems as well as model-based decision control tools. These tools use historical process data to create statistical as well as machine-learning based prediction models.

The scrap mix optimisation system is applied to select the cost-optimal mix of scrap types and alternative charge materials for production of each product quality. This also includes the influence of different charge material on the energy demand of the process (c.f. Carlsson et al [6]). The model also includes the current availability of the different charge material types into the optimisation. Besides the classic optimisation methods, a set of innovative data analytics tools is implemented to highlight correlations between inputs (scrap quality) and outputs (product quality).

The energy efficiency optimisation system is based on the use of machine learning techniques. A data-driven modelling approach is used to identify the patterns that represent normal process behaviour from previous feedstocks. The data from previous material and energy inputs is used to forecast the expected energy consumption (c.f. Manojlović et al [7]). Once the appropriate statistical models are analysed, learned and sufficiently trained, the energy efficiency is derived from it. The generation of this model trained with the aforementioned data allows forecasting future hypothetical situations that can be reached when a new combination of parameters is proposed.

And lastly, the model-based decision control tools are used to monitor and predict process behaviour during melting. The systems are mainly based on mathematical models using energy and mass balance calculations. Such dynamic energy and mass balance models allow the continuous calculation of the melt temperature based on the specific meltdown energy demand of the charged materials and the electrical and chemical energy input. With modelling of the metallurgical reactions which are performed during the melting and refining process, also the chemical composition of the metal melt can be calculated continuously. More detailed information on the work on the retrofitting solutions can be found on the project website (www.revamp-project.eu [8]), where all non-confidential reports published within the project are available.

3. Use Case: Electric steelmaking

In electric steelmaking up to 100 % of the metallic charge materials for the EAF is recycled steel scrap. Figure 1 shows a typical process route for electric steelmaking, as it is used at the steel plant of Sidenor, with indication of the methodologies and tools to be applied within the demonstration use case.

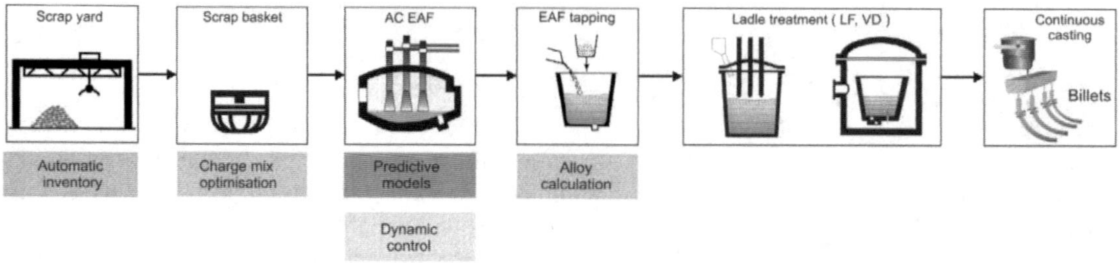

Figure 1. Liquid steelmaking process route of the electric steelmaking plant at Sidenor

The incoming scrap is delivered by trucks and is sorted within dedicated boxes and areas at a scrap yard, according to the delivered scrap types. In the scrap yard, the individual scrap mix for producing one batch for a certain steel grade is loaded by cranes into two scrap baskets, which are used to charge the scrap into the furnace. For on-line monitoring and control of the EAF melting and refining process, predictive process models and dynamic control set-point calculations regarding material and energy input were applied, in order to achieve the target end-point of the process with respect to melt temperature and composition under minimised energy and material input. After melting and refining in the EAF, the liquid steel melt is tapped into a ladle, and in parallel first deoxidation and alloy materials as well as slag formers are added. Their amounts were calculated depending on the calculated temperature and composition of the steel melt tapped from the EAF. During the following ladle treatment, the steel melt is further refined at a Ladle Furnace (LF) and a Vacuum Degassing (VD) station. Finally, the liquid steel melt is casted and solidified within a continuous casting plant to produce steel billets and blooms.

An overview of the analysed process from the steel use case in the project is shown in the flowchart below (Figure 2). Shown here are all analysed processes and the considered input and output flows.

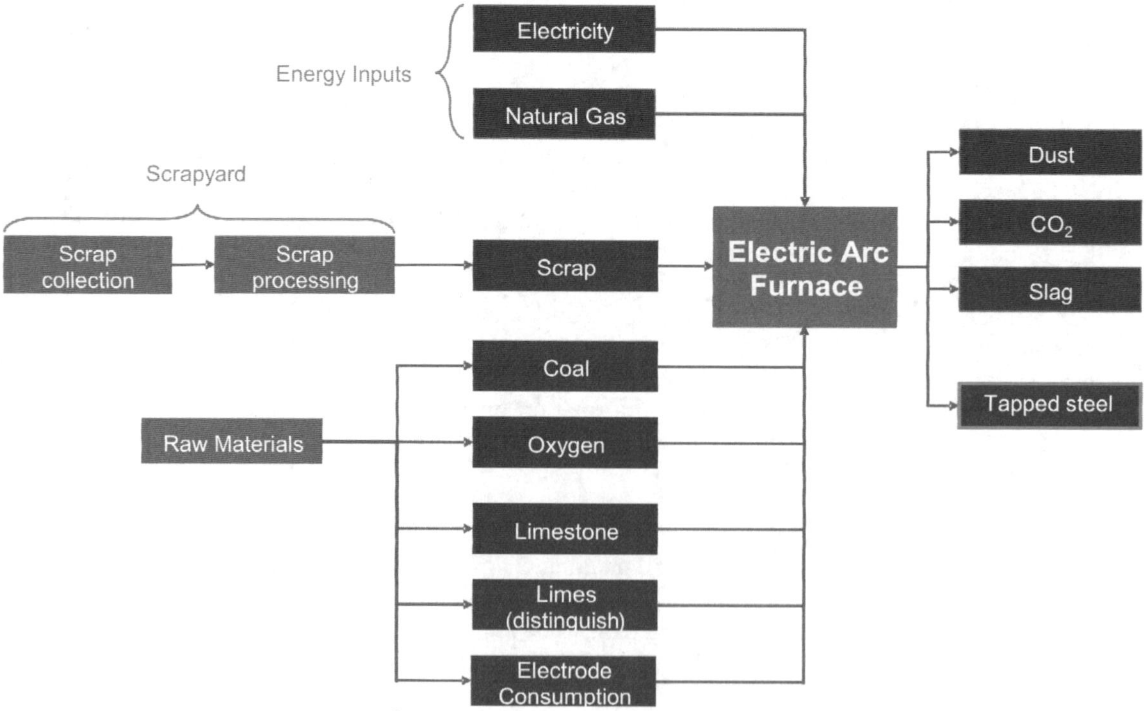

Figure 2. Flowchart of the steel production through an Electric Arc Furnace

The electric energy input combines the required input for the scrapyard operations and for smelting operation. The off-gas was simplified to the dust and CO_2 components. Based on the flowchart, the input and output parameters have been determined with data according to Sidenor's usual production process and are implemented in the model. Adequate upstream processes for resource supply were modelled using the ecoinvent® database. The oxygen used in the EAF is produced on site via cryogenic air separation. Regarding electric energy, the specific electricity mix of Sidenor's provider was implemented as the source of all electric energy. Based on this flowchart, the input and output quantities of the various parameters were collected. A table with the used process data is shown in Table 1a and Table 1b.

Table 1a. Input Quantities used in the production process at Sidenor

Parameter	Value	Unit
Electric energy	458.6	kWh/t_{Steel}
Natural gas	1.87	Nm^3/t_{Steel}
Scrap	1.13	kg/t_{Steel}
Coal	15.37	kg/t_{Steel}
Oxygen	22.99	Nm^3/t_{Steel}
Limestone	32.2	kg/t_{Steel}
Lime (injected)	2.8	kg/t_{Steel}
Electrode Consumption	2	kg/t_{Steel}

Table 1b. Output Quantities used in the production process at Sidenor

Parameter	Value	Unit
Tapped steel	1	t/t_{Steel}
Slag	150	kg/t_{Steel}
Dust	20	kg/t_{Steel}
CO_2	50	kg/t_{Steel}

The performance of the retrofitting solutions is evaluated via a Life Cycle Analysis (LCA). An LCA aims to evaluate environmental impacts associated with a product. The exact procedure is specified by the international standards ISO 14040 [4] and 14044 [3]. An LCA is divided into the four phases "Definition of goal and scope", "Life Cycle Inventory", "Impact assessment", and "Interpretation". These phases are carried out iteratively and influence each other. [2]

Phase 1 focuses on defining the framework of the LCA. The intended application and audience are especially considered, as well as necessary assumptions. In phase 2, the life cycle inventory, an overview of the input and output data of the analyzed system is established. This includes data collection, calculations and if necessary, allocation. The life cycle impact assessment (LCIA) in phase 3 takes place based on the results of the life cycle inventory and derives the environmental impacts caused from the input and output data of the life cycle inventory. For this purpose, the input and output data of the life cycle inventory are weighted and aggregated in various impact categories and indicators. However, the impact assessment only identifies potential environmental impacts based on estimations. In the fourth and final phase, the results of the life cycle inventory and impact assessment are evaluated in relation to the initial goal of the LCA. Due to their complexity, life cycle assessments are typically performed using software solutions in combination with extensive generic databases. In this case, the Umberto® LCA+ software is used in combination with the ecoinvent® database [5]. [2]

The results are related to the functional unit of 1 t of liquid steel, tapped at the end of the electric arc furnace process. Since no cyclical processes are considered and a holistic analysis (cradle-to-grave) is not taken, no allocation procedures for by-products are considered. Input flows and by-products that leave the system boundary are considered by means of system area expansion. The impact assessment methodology used is the ReCiPe-Midpoint Method combined with data from the ecoinvent® database.

The selected environmental LCIA indicators used to estimate the environmental impacts using the calculated inputs and outputs are the following:

- Global Warming Potential, GWP
- Fossil Depletion Potential, FDP
- Freshwater Eutrophication Potential, FEP
- Metal Depletion Potential, MDP
- Natural Land Transformation, NLTP
- Ozone Depletion Potential, ODP
- Photochemical Oxidant Formation Potential, POFP
- Terrestrial Acidification; TAP
- Human Toxicity Potential, HTP
- Water Depletion Potential, WDP

These indicators cover all major impact categories specified by the norms. The values of the respective indicators are determined using LCA software. It should again be emphasized that the impact assessment is not equivalent to a toxicological risk analysis and only identifies potential environmental effects [3].

The meaning of the impact indicators and in particular their units are explained below separately. Due to the large number of substances that have an influence on the level of an impact indicator, one (usually mainly relevant) substance is selected for each impact category. The effect of all other substances of this impact category is then set in relation to the selected substance. An example of such a calculation is methane (CH_4) in the calculation of the Global Warming Potential (GWP) impact indicator. The global warming potential is usually given in kg CO_2-eq. According to the latest IPCC publications [1] the impact of methane on the greenhouse gas effect is approx. 300-times as large as the impact of carbon dioxide. Similar comparisons are made for all substances in all impact categories. It should be noted though, that the underlying selection of these reference substances is arbitrary. However, the units of the impact categories selected in this contribution have been generally established at the international level. [2]

First, data from the state-of-the-art production process is recorded as a base line. The use case resembles the usual production process carried out by the industrial partner. The model focuses solely on the process step taking place at the Electric Arc Furnace and before that. The secondary metallurgy in the ladle is not considered. This base line is then compared against an altered state of the process, where the retrofit solutions are installed. Among the retrofitting solutions, the scrap mix optimizer is already assessing the optimal scrap mix, the dynamic control process has been proven for an on-line assessment. The energy efficiency optimization model will be finalized until the end of the project. Thus, the following changes in process parameters have been determined based on the evaluation of the different solutions:

- Reduction of electric energy input (2-4%) and natural gas input (3-5%)
- Less scrap needed (1-3%) due to better sorting
- Less injection coal and oxygen (1-3% each) needed

The calculations are conducted accordingly to cover the full range of potential changes described above. These process improvements only cover the input side of the process described above, but no changes of the output parameters (mainly off-gas data) is considered. In reality it is to be expected, that these modifications also result in changes to the off-gas parameters of the furnace. However, these changes are difficult to foresee right now, and therefore no estimations in this regard are made. As a result, some impact indicator changes are underestimated, which is pointed out at the corresponding points.

4. Results

The process model created within the Umberto® LCA+ software is shown in Figure 3 below. Through the Sankey Diagram function the largest material and energy flows of the system are highlighted, which are in this case the input scrap, the electric energy and the product output.

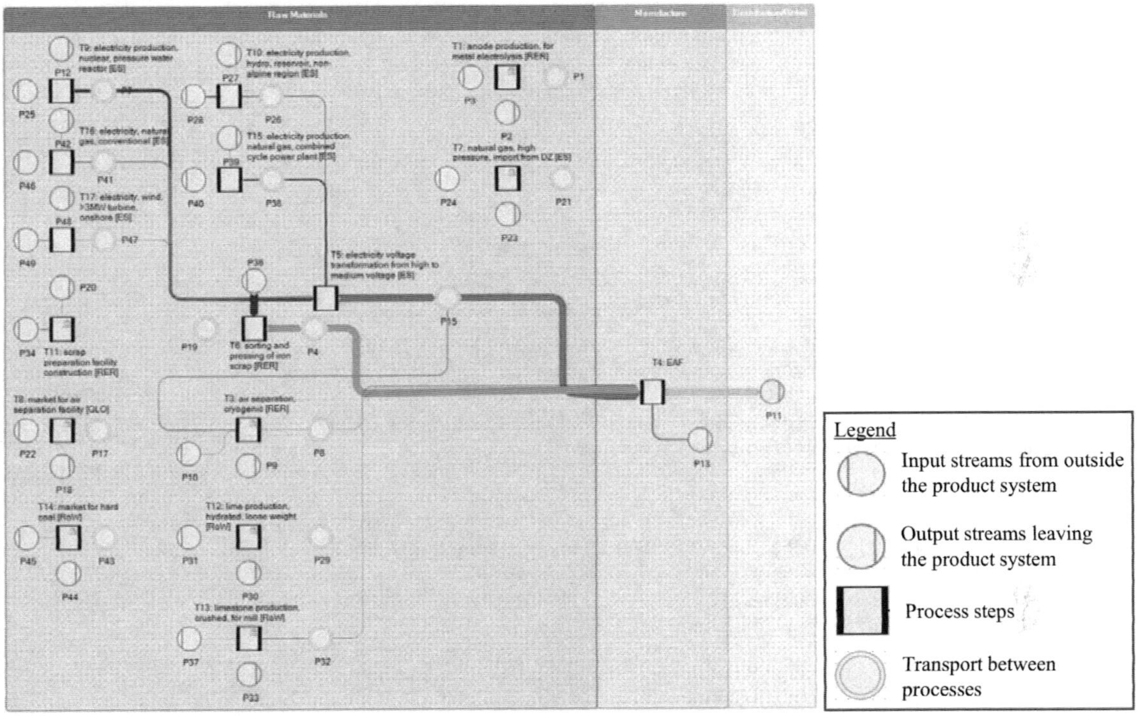

Figure 3. Process model in Umberto® LCA+ with a Sankey Diagram

To more accurately reflect the specific use case of the industrial partner, the specific electricity mix of the industrial partner is reproduced in the model according to the information provided by the grid operator. For this reason, the electricity generation in the model shown below is broken down to the individual power generation methods. The results for the specified LCIA indicators are shown in Figure 4 below.

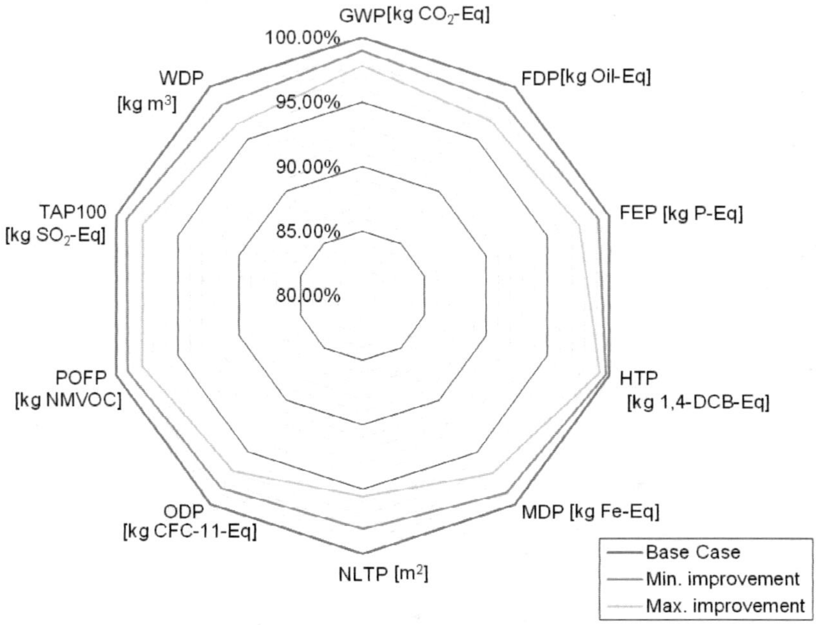

Figure 4. Relative change in selected LCIA indicators due to estimated process improvements

The presentation of the results takes place only relatively between the different states, since concrete numerical results cannot be shown for confidentiality reasons. The underlying perceptual values are given in Table 2.

Table 2. Percentual values of LCIA indicator changes

Parameter	Base Case	Min. improvement	Max. improvement
GWP	100.00%	98.97%	97.81%
FDP	100.00%	98.49%	96.78%
FEP	100.00%	99.15%	97.58%
HTP	100.00%	99.74%	99.25%
MDP	100.00%	98.89%	97.05%
NLTP	100.00%	98.09%	95.57%
ODP	100.00%	98.51%	96.88%
POFP	100.00%	99.07%	97.87%
TAP	100.00%	99.15%	97.83%
WDP	100.00%	98.31%	96.44%

The biggest indicator changes resulting from the process modifications were recorded for the Fossil Depletion Potential FDP, the water depletion potential WDP and the natural land transformation NLTP. These improvements result largely from the possible natural gas and electric energy savings accomplishments. In general, it was observed in all impact categories that energy use and thus energy efficiency measures have the highest influence on the ecological evaluation. The human toxicity potential shows little change due to the modifications, with 0.26 % up to 0.75 % improvements. This is due to the fact, that the biggest contributor to this impact category are the changes of the off-gas being emitted from the electric arc furnace, which is assumed as constant throughout the process modifications. But as explained before, this circumstance is expected to differ in reality once the retrofitting solutions have been installed. These limitations must also be considered when interpreting the results of the global warming potential (GWP), photochemical oxidant formation potential (POFP), and terrestrial acidification (TAP) categories, with the influence being highest for the GWP. In these categories as well, higher changes in the indicators are to be expected in reality than estimated here.

5. Discussion

The impact assessment of the altered process has been stated based on the results obtained by the retrofitting solutions taking into account their influence on the production process. However, the influence of the process modifications to the output side of the process, especially the off-gas, can't be foreseen for now, and are therefore assumed as static in these analyses. The impact in reality will most likely be different than estimated here.

Additionally, the process has been simplified in this case to the most relevant input and output parameters. This selection was made in cooperation with the industrial partner to ensure completeness of the model. Material or energy flows were only cut-off, if the volume and the influence on one or more impact indicators was insignificant. The upstream process data for material and energy supply was selected with regard to the circumstances of the real facility, and gaps were closed with process data from the ecoinvent® database specific for the geographical location of the plant (in this case Spain).

6. Conclusion

The analyses show, that material and energy efficiency measures have a large impact on the ecological evaluation of the process. The described retrofitting solutions target both of these areas and can contribute to lower the ecological impact of the steel recycling process. The described retrofitting solutions influence the ecological impact indicators and decrease their values by up to 4.43 %, considering the assumptions taken. The concrete figures are expected to change for some impact indicators, once enough process data from the daily production using the new technologies is gathered and analysed.

7. Acknowledgement

The research leading to these results has received funding from the European Union's Horizon 2020 research and innovation programme under grant agreement ID 869882.

References

[1] IPCC: "Climate Change 2013: The Physical Science Basis. Contribution of Working Group I to the Fifth Assessment Report of the Intergovernmental Panel on Climate Change, Stocker, T.F., D. Qin, G.-K. Plattner, M. Tignor, S.K. Allen, J. Boschung, A. Nauels, Y. Xia, V. Bex and P.M. Midgley (2013)Another reference

[2] Klöpffer, W.; Grahl, B.: "Ökobilanz (LCA)", Wiley-VCH, 2009

[3] ISO 14044, Berlin (May 2018)

[4] ISO 14040, Berlin (November 2021)

[5] Wernet, G., et al.: "The ecoinvent database version 3 (part I): overview and methodology", The International Journal of Life Cycle Assessment, pp. 1218–1230 (2016)

[6] Carlsson, L. S.; Samuelsson, P. B.; Jönsson, P. G.; "Modeling the Effect of Scrap on the Electrical Energy Consumption of an Electric Arc Furnace", Processess, pp. 1044 (2020)

[7] Manojlović, V.; Kamberović, Ž.; Korać, M.; Dotlić, M.: "Machine learning analysis of electric arc furnace process for the evaluation of energy efficiency parameters", Applied Energy, Vol 307 (2022)

[8] VDEh-Betriebsforschungsinstitut GmbH (BFI): "Official project website of the REVaMP (Retrofitting Equipment for Efficient Use of Variable Feedstock in Metal Making Processes) project", URL: https://www.revamp-project.eu, 22.01.2024

Biochar as a slag foaming agent in EAF – A novel experimental setup

E Hoikkaniemi[1,3], P Sulasalmi[2], V-V Visuri[2] and T Fabritius[2]

[1] SSAB Europe Oy, PO Box 93, 92101 Raahe, Finland
[2] Process Metallurgy Research Unit, University of Oulu, PO Box 4300, 90014 University of Oulu, Finland
[3] Author to whom any correspondence should be addressed.

eetu.hoikkaniemi@ssab.com

Abstract. Slag foaming practice is employed widely in electric arc furnace steelmaking to improve the energy efficiency, protect the furnace structures and reduce noise pollution. Slag foaming is typically launched by injecting fossil-based carbonaceous material (e.g. coke dust) into the melt. In this study, a novel laboratory-scale experimental setup was used for studying the substitution of fossil carbon by biochar as slag foaming agent. The setup was equipped with an injection device for feeding carbon and a camera system for observing the foaming phenomenon and recording the process. A set of experiments was conducted for studying the foaming in slag-carbon systems using two carbonaceous materials: 1) coke dust was used as a reference material and 2) a high-quality biochar was used as a possible replacement. In the experiments, sufficient foaming was achieved with both of the carbonaceous materials. The biochar produced almost equal foaming behavior as coke dust. The results indicate that biochar could be used to substitute carbon for slag foaming in the EAF.

1. Introduction

The foaming of slag plays an important role in modern electric arc furnace (EAF) steelmaking and has several positive effects on the process. It improves the energy efficiency of the process, reduces noise pollution, inhibits steel re-oxidation and enhances the service life of the refractory, wall panels and electrodes of EAF [1–3]. Nowadays, the foaming of slag in EAF is usually generated by injecting oxygen and carbonaceous material from fossil sources, e.g. coke or anthracite, into the steel-slag interface, starting from the point of the process when the scrap charge is nearly melted. The carbonaceous material reacts with FeO in the slag producing CO gas bubbles. These CO gas bubbles mainly cause the foaming of the slag [4]. The typical amount of injected carbon for slag foaming purposes is 5–10 kg/ton of steel, which results in CO_2 emissions of 24.2 kg/ton of steel, on average [5]. In order to move to fossil-free steelmaking, alternatives to those fossil carbon-based foaming agents must be found; the slag foaming is a critical part of the process, and it cannot be completely left out of the process or replaced. Biochar is considered to be a potential material for replacing the fossil foaming agents [6]. Replacing fossil-based carbon with biochar has a potential of reducing direct CO_2 emissions of EAF process by 25%.

The purpose of this study was to study the suitability of biochar for slag foaming. The study was made at a laboratory-scale, using a novel experimental setup where laser light source-based illumination system was used to observe the EAF slag foaming phenomenon. Coke dust was used as a reference material, and a high-quality biochar was used as a bio-based material in the experiments.

2. Materials and methods

The slag used in the experiments was obtained from industrial EAF process. The lumpy slag collected after cooling was ground to <1 mm particle size to suit the purposes of the foaming experiments. To adjust the basicity of the slag to a more suitable value, reagent grade CaO was added to the slag. Based on previously made studies [7–9], the basicity value (CaO/SiO_2) was adjusted to 2.0. The chemical composition of the slag used in the experiments can be seen in table 1.

Table 1. The chemical composition of the slag used in the foaming experiments.

Content (wt-%)								
CaO	SiO_2	Al_2O_3	MgO	MnO	TiO_2	V_2O_5	FeO	CaO/SiO_2
33.34	16.59	7.06	7.72	0.85	1.08	1.15	29.02	2.0

The coke dust was obtained from an industrial coking plant and the biochar was obtained from a company that manufactures biochar with industrial scale pyrolysis equipment. The pyrolysis temperature was 800 °C and the pyrolysis time was roughly 10 minutes. The biochar was made from spruce wood chips, so it must be ground prior the experiments to obtain a particle size close to that of the coke dust. The grinding was made with a Retsch ZM 200 centrifugal mill. The chemical composition of the coke dust and the biochar used in the experiments can be seen in table 2. Since the moisture content of the ground biochar was as high as 11.84 wt-%, it was dried 24 h at 105 °C prior the experiments to avoid the effect of moisture on the foaming.

Table 2. The chemical composition of the coke dust and the biochar used in the experiments.

Material	Chemical analysis (wt-%)						Ash composition (wt-%)					
	Fixed C (db[a])	Volatile matter (db[a])	Ash (db[a])	Moisture	S	P	SiO_2	CaO	Al_2O_3	MgO	K_2O	Na_2O
Coke dust	84.41	0.95	14.64	0.10	0.85	0.07	7.10	0.89	4.10	0.19	0.22	0.05
Biochar	92.72	4.65	2.63	11.84	<0.03	0.10	0.15	1.05	–	0.14	0.17	0.02

[a] db = dry basis.

The foaming experiments were conducted in a Nabertherm HT 08/18 chamber furnace (chamber dimensions 150x150x300 mm, maximum temperature 1800 °C). The interior temperature of the furnace was measured with a thermocouple placed to the roof of the furnace. Ground EAF slag mixed with the reagent grade CaO was loaded to a crucible (inner diameter 58 mm, inner height 80 mm) which was made by cutting and drilling from MgO-C refractory brick. The crucibles were pre-burned before the experiments in 1000 °C to reduce the amount of carbon in the surface parts of the crucibles. The crucible loaded with the slag was placed into the chamber furnace and slightly tilted to handle the possibly overflowing slag. The furnace was heated with a 2 °C/min heating rate to 1400 °C, which was the foaming experiment temperature. When the temperature was reached and the slag was in a liquid form, carbonaceous material was injected into the slag via a quartz pipe lance with inner diameter of 4 mm. Nitrogen gas was used as an inert carrier gas for the carbonaceous material with a 0.15 Nl/min flow rate. The carbonaceous material was fed to the gas line with a screw feeder with different feeding rates.

A schematic of the experimental setup can be seen in figure 1. The injection period was filmed with a visualization system which consisted of a CAVILUX HF laser light source system connected to a Baumer TXG14 CCD camera. This system was operated with a computer, and as a result of the visualization clear video image of the process was obtained. The laser and the camera were aimed to the crucible through a hole drilled to the roof of the furnace. During the heating period, the hole was covered with a lid, which was then removed just before the injection begun. A total of four foaming experiments were made as a part of this set of experiments; two reference experiments with the coke dust and two experiments with the biochar. An overview of the parameters used in the experiments can be seen in table 3.

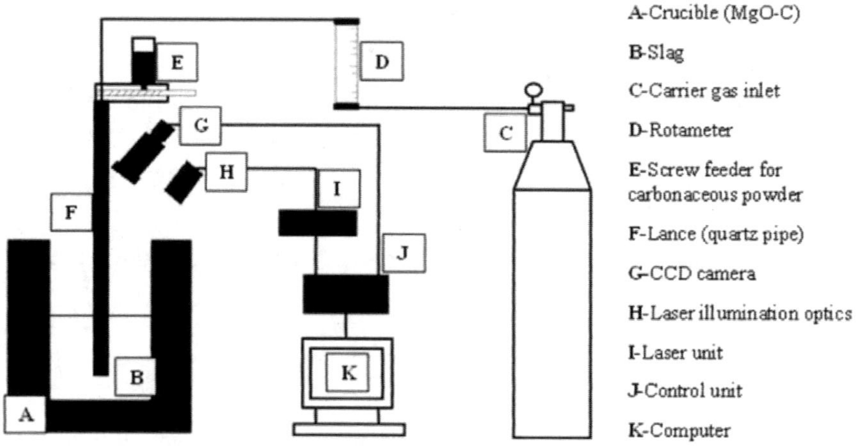

Figure 1. A schematic of the experimental setup.

Table 3. Overview of parameters used in the foaming experiments.

Exp. No.	Foaming agent	T (°C)	W_{slag} (g)	$Q_{carrier\ gas}$ (Nl/min)	$t_{injection}$ (min)
1	Coke dust	1400	400	0.15	4
2	Coke dust	1400	300	0.15	4
3	Biochar	1400	400	0.15	9.5
4	Biochar	1400	300	0.15	4

3. Results

The results of the slag foaming experiments can be divided into two parts; the visual foaming phenomenon analysed from the video frames, and the microscopic examinations done to the slag after the experiments.

The two reference experiments done by using coke dust as a foaming agent proved that the experimental setup was usable and that the foaming of the slag could be recorded with the video camera. When the injection lance was lowered into the slag, only nitrogen gas was injected during the first seconds. This did not cause a significant rise of the surface of the slag. When the injection of the coke dust was turned on, the slag surface started to rise immediately. The rise of the slag surface was mostly constant, but at times, the surface level oscillated slowly. In both of the reference experiments, two separate 2-minute injection periods were conducted, and between those, the slag surface was allowed to settle by injecting only nitrogen gas into the slag. In both experiments, the first injection period was more effective than the latter one.

The two experiments made with the biochar were performed in such a way that they were comparable to the reference experiments. The only exception was that in the first experiment, the total injection time was 9.5 minutes, which was divided into four short, roughly 2-minute periods of injection. In the second experiment, the 4-minute total injection time divided into two 2-minute injection periods was implemented. In both of the experiments, the rise of the surface of the slag was notable. In the first experiment, the slag surface rose slightly less than in the reference experiments, but in the second experiment, the surface rose similar as in the reference experiments. The highest point of the slag was estimated to be even 10 mm higher than in the corresponding experiment made with the coke dust, and the slag did overflow from the crucible.

In figure 2, an example image from foaming experiment made with the coke dust can be seen; the zero-level of the slag is on the left and the highest point of the slag surface is on the right. The measurement scale in the middle was made so that a ruler was placed into the empty crucible prior the experiment. This was done separately for all of the experiments, and with this method, a good estimate from the change of the height of the slag bed could be made, even though the values are not completely accurate. Table 4 provides a compilation of the results regarding the height of the slag.

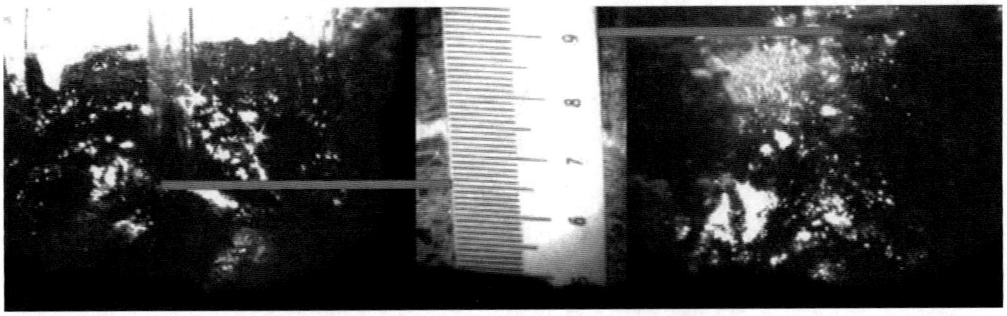

Figure 2. The zero-level and the highest point of the slag surface in a reference experiment.

Table 4. Results of the foaming experiments.

Exp. No.	Foaming agent	Slag					Injection	
		h_0[a] (mm)	Δh[a] (mm)[b]	t_{rise}[a] (s)[b]	t_{fall}[a] (s)[b]	ΔV[a] (%)[b]	W_{agent}[a] (g)	Avg. rate (g/min)
1	Coke dust	65	25[c]	85	80	38.5	1.54	0.39
2	Coke dust	48	40[c]	75	90	83.3	3.00	0.75
3	Biochar	65	16	25	30	24.6	6.66	0.70
4	Biochar	48	50[c]	42	110	104.1	5.85	1.46

[a] h_0 = zero level of the slag, Δh = change of the slag height, t_{rise} = slag surface rise time, t_{fall} = slag surface fall time, ΔV = change of the slag volume, W_{agent} = total amount of material injected.
[b] Results based on the injection period where the highest point of the slag surface was obtained.
[c] The highest point was roughly the top edge of the crucible.

As seen from table 4, the rise time of the slag was considerably faster in the experiments made with the biochar compared to the reference experiments. The differences may be attributable to possible differences in the injection rates of the materials at the beginning of the injection. The flow behavior of coke dust and biochar differed, and since the feed of the material was made manually without any kind of control mechanism, the flow rate could not be reliably adjusted. This is one thing that could be developed in further research. When looking at the values for the change of the volume of the slag during the foaming, it can be concluded that both materials induced slag foaming, and exhibited no clear differences in foaming efficiency.

After the experiments, the slag was allowed to cool down slowly in air atmosphere inside the crucibles. The crucibles were visually analysed after they had cooled to room temperature. Some splashes of mostly reduced iron could be observed on the top-edges of the crucibles. The crucibles were split from the middle with a saw, and inside the slag bed, especially near the bottoms of the crucibles, some reduced iron could be seen in both of the cases, the reference experiments and the biochar experiments. From each of the split crucibles two sections of slag with a diameter of roughly 35–40 mm were cut for microsection sample preparation. The prepared microsection samples were then studied using digital light optical microscope (Olympus DSX1000) and FESEM (Zeiss Sigma) equipped with EDS detector. With digital microscope, a closer look was taken to the shiny droplets identified as reduced iron. These droplets were observed inside the bulk slag in all the microsection samples except the ones made from the slag of the experiment No. 3. Most of the droplets were roughly 10–100 μm in sizes, gathered together in either "cloudy" type or long and thin formations, and some of the droplets were supported by air bubbles that had remained inside the solidified slag. Using EDS point analysis, the light-colored droplets could be confirmed to be reduced iron. Figures 3 and 4 present examples of the digital microscope images and FESEM-EDS analysis done for the slag after the foaming experiments. No quantitative analysis was made for the reduced iron in the samples, so it is difficult to evaluate the reduction efficiency between the coke dust and the biochar.

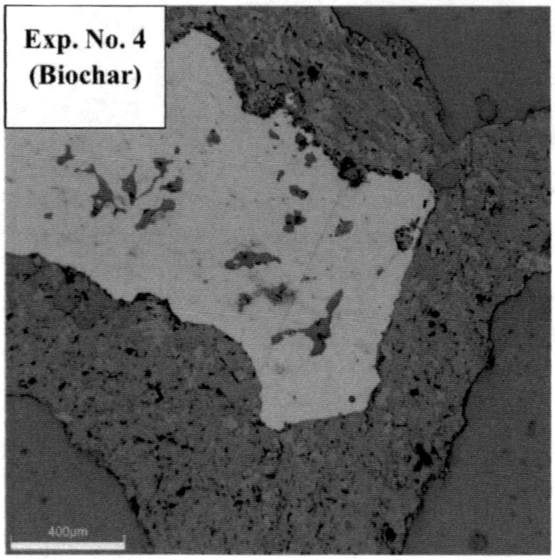

Figure 3. Digital microscope image taken from the slag after the experiment No. 4.

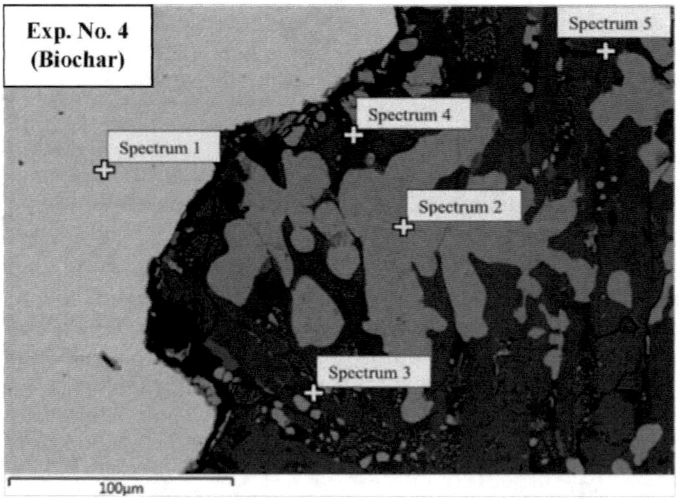

Spectr.	O	Na	Mg	Al	Si	P	Ca	V	Cr	Mn	Fe	W
1					0.13						99.87	
2	21.80		7.21					0.59	0.24	1.50	68.32	0.33
3	36.47	0.25	5.78		16.57	0.30	37.12	0.27			3.23	
4	39.36		0.77	16.41	11.00		29.93				2.53	
5	36.00		2.93		16.43		42.52	0.26			1.87	

Figure 4. The results of the FESEM-EDS analysis done for the slag after the experiment No. 4.

4. Conclusions

A novel experimental method was built to study the slag foaming phenomenon on a laboratory scale. With the method, it is possible to record image from the foaming phenomenon and to accurately see how the injection of materials into the slag changes the volume of the slag. The method gives better understanding on the foaming phenomenon, and is suitable for studying e.g. the differences in efficiency between different slag foaming agents. A total of four slag foaming experiments were made as a part of this study; two reference experiments with coke dust, and two experiments with high-quality biochar.

The results showed that biochar is a potential material to replace fossil-based foaming agents. Based on the video material obtained, the biochar acted as a foaming agent in a similar way than the coke dust. The surface of the slag rose with both of the materials in a same kind of manner, and there appeared no clear difference between the biochar used and coke dust in terms of the foaming efficiency. The FESEM-EDS analysis conducted for the cooled slag after the experiments proved that both carbonaceous materials injected into the slag clearly reduced iron oxide of the slag to metallic iron, thereby creating CO gas that makes the slag foam.

The study partly confirms the observations made at previously made studies. However, a large part of the results in the previously made studies has been the unclear functionality of the biochar; some of the materials tested have showed clear potential to foam the slag, but some of the materials have been considered to be unreactive. Thus, it cannot be generally concluded that all sorts of biochar are suitable foaming agents.

Acknowledgments
This study was executed within the Towards Fossil Free Steel (FFS) project funded by Business Unit Finland.

References

[1] Luz A P, Ávila T A, Bonadia P and Pandolfelli V C 2011 Slag foaming: fundamentals, experimental evaluation and application in the steelmaking industry *Refractories WorldForum* **3** 91–8
[2] Chang Y E, Lin C M, Shen J M, Chang W T and Wu W 2021 Effect of MgO content on the viscosity, foaming life, and bonding in liquid for the CaO-SiO2-MgO-5Al2O3-30FeO slags *Metals* **11** 1–12
[3] Zhu T X, Coley K S and Irons G A 2012 Progress in slag foaming in metallurgical processes *Metall. Mater. Trans. B* **43** 751–7
[4] Corbari R, Matsuura H, Halder S, Walker M and Fruehan R J 2009 Foaming and the rate of the carbon-iron oxide reaction in slag *Metall. Mater. Trans. B* **40** 940–8
[5] Thomson M J, Evenson E J, Kempe M J and Goodfellow H D 2000 Control of greenhouse gas emissions from electric arc furnace steelmaking: evaluation methodology with case studies *Ironmak. Steelmak.* **27** 273–9
[6] Bianco L et al. 2013 Sustainable electric arc furnace steel production: GREENEAF *Berg-hüttenmänn. Monatsh.* **158** 17–23
[7] Ito K and Fruehan R J 1989 Study on the foaming of CaO-SiO2-FeO slags: part I. foaming parameters and experimental results *Metall. Mater. Trans. B* **20** 509–14
[8] Ito K and Fruehan R J 1989 Study on the foaming of CaO-SiO2-FeO slags: part II. dimensional analysis and foaming in iron and steelmaking processes *Metall. Mater. Trans. B* **20** 515–21
[9] Jung S M and Fruehan R J 2000 Foaming characteristics of BOF slags *ISIJ Int.* **40** 348–55

Application of hydrogen operated burners in the electric arc furnace

L Schüttensack, A Reinicke and T Echterhof

Department for Industrial Furnaces and Heat Engineering, RWTH Aachen University, Kopernikusstr. 10, 52074 Aachen, Germany

schuettensack@iob.rwth-aachen.de

Abstract. The steel making industry has one of the highest carbon dioxide emissions worldwide. The recycling route with the electric arc furnace (EAF) as the central aggregate emits less carbon dioxide in comparison to the blast furnace. Still, the EAF offers high potential for a reduction of the carbon dioxide emissions. One possibility is the substitution of natural gas by hydrogen in the auxiliary burners of the EAF. This substitution is investigated and analysed using a dynamic EAF process model. The model is able to simulate the combustion of natural gas as well as hydrogen. The differences in the operating conditions and properties by the hydrogen operated burners can be analysed in the results of the EAF process model.

1. Introduction

The electric arc furnace (EAF) is the most used aggregate in the steelmaking industry for steel recycling and one of the most used aggregates for the primary steel production [1]. In comparison with other aggregates for steel making e.g. the blast furnace, the EAF is low in greenhouse gas emissions [2]. The main energy source in the EAF is electric energy which results in lower direct emissions. Still the carbon footprint for a ton of steel is high and leaves room for improvements.

In addition, auxiliary burners are utilized in the operation of the EAF. These are used as a secondary energy input and applied to heat "cold spots" in the steel scrap. The burners are operated with natural gas and are therefore a source of carbon dioxide in the EAF operation. The substitution of natural gas by hydrogen eliminates the carbon dioxide emissions from the burners and is therefore a way to further reduce the greenhouse gas emissions of the EAF. The combustion of hydrogen influences all processes in the EAF and thus the effect has to be analysed.

A dynamic EAF process model first developed by Logar et al. [3]-[7] and implemented and improved by Meier [8] and Hay [9]-[12] can be used to simulate the hydrogen combustion and the effects on the EAF operation. The dynamic EAF process model is able to simulate the complex processes inside the EAF.

In the literature a variety of process models for the EAF is available. Due to the high complexity of the operation most models differ in the assumptions made to simplify the model. A detailed modelling of the burners is often not considered or described and the output of the complete combustion of natural gas is taken as an input for these models [13]. However, MacRosty et al. [14, 15] considers the burners in the gas zone of their model. Methane and oxygen are fed via the burners to the gas zone, where the combustion of methane is taking place, leading to the state of chemical equilibrium in the gas zone. Another approach, used by Rathaba [16], takes the combustion with a mixed fuel gas into account, including hydrogen, methane and other species. The efficiency of mixing and combustion is considered using an adjustable parameter, resulting in a not complete combustion.

In scope of the RFCS funded project "DevH2forEAF" the substitution of natural gas by hydrogen in the auxiliary burners of the EAF is investigated. The effects of the hydrogen combustion on the EAF operation are simulated using the EAF process model with the addition of the hydrogen combustion to the process model. For further investigation of the effects of the hydrogen combustion on the EAF process, a pilot scale burner is operated within a pilot scale EAF. Especially the effect on the melt and slag composition are analysed. These findings will help with the adjustment and validation of the hydrogen combustion in the EAF process model.

2. Materials and methods

The EAF is a highly complex system with a lot of variety in the operation parameters such as the scrap composition. Furthermore, the extreme conditions within in the EAF hinder the direct measurement of important properties during operation, for instance the temperature of the steel melt. The usage of a model simulating the EAF process can overcome this problem and give an insight on the processes inside the EAF.

The dynamic EAF process model consists of eight zones, as shown in Figure 1. Each zone has a homogeneous composition and temperature. The changes in composition and temperature of each zone are modelled by application of a mass and energy balances. The electric arc as well as the refractory lined walls and water-cooled vessel are respectively considered as a heat source or sink. These heat sources and sinks do not have a separate mass and energy balance in the model. Heat is transferred between the zones via direct contact as well as radiation, in addition chemical reactions and phase changes are also considered in the model. With this the resulting change of composition and temperature in each zone can be calculated and is reflected in the mass and energy balance of the zone.

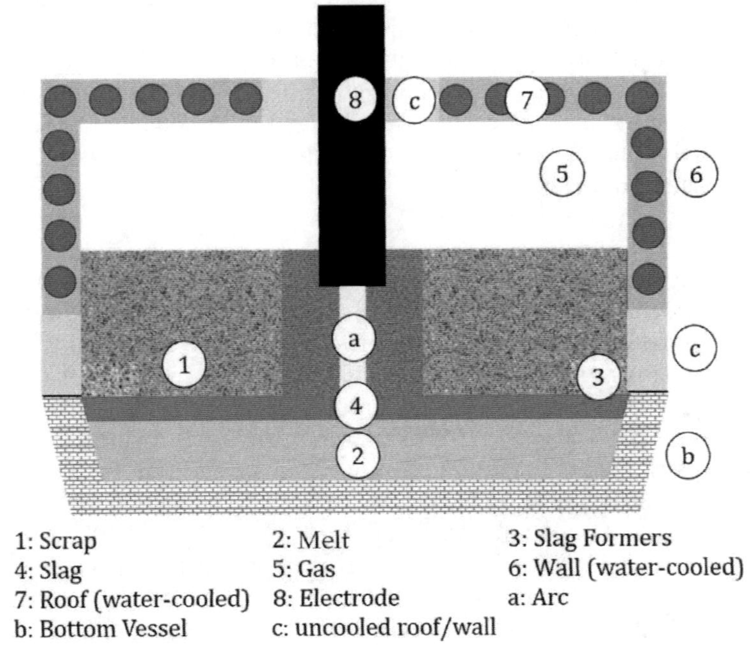

Figure 1. Model zones (1-8) and heat sources/sinks (a-c) in the EAF process model. [12]

IOP Conf. Series: Materials Science and Engineering 1309 (2024) 012011 doi:10.1088/1757-899X/1309/1/012011

For the estimation of the influence of hydrogen fired burners on the EAF process, the model is further expanded to simulate the hydrogen combustion. As part of the gas zone, the reactions of the burners installed in the EAF are considered. Since hydrogen is already present as a species in the gas phase, only the primary hydrogen combustion reaction is added to the model (equation (1)).

$$H_2 + 0.5\ O_2 \rightleftharpoons H_2O \quad (1)$$

With the addition of this equation and the corresponding parameters and considerations, it is possible to simulate burners operating with different fuel gas compositions. These fuel gas compositions can range from 100% natural gas to 100% hydrogen, including all mixtures in between. The substitution of natural gas by hydrogen primarily effects the burner reactions taking place. As such the fuel gas composition determines the gas composition within the furnace. Ultimately, this affects the chemical reactions taking place at the surface of the scrap, melt and slag as well as the radiative heat transfer within the furnace. This is particularly the case as CO, CO_2 and H_2O are all relevant species in regards to the absorption of radiation of the gas phase. This way, the overall energy distribution within the furnace changes even when the energy introduced into the furnace by the burners is kept constant. This effect is represented by the results of the EAF process model. For a comparison of the operation of natural gas and hydrogen fired burners, parameters for the substitution of natural gas by hydrogen have to be determined. The caloric value of the burners has to be constant to generate the same energy input trough the burners independent of the fuel gas. Constants can be calculated to adjust the mass flow of natural gas and oxygen of the burner, which results in a mass flow of hydrogen and oxygen with the same caloric value. However, the results of the EAF process model for the combustion of hydrogen can not be validated yet. This validation will be carried out after the industrial trials with hydrogen burners.

As a part of the RFCS funded project "DevH2forEAF" the substitution of natural gas by hydrogen in the auxiliary burners of the EAF is investigated. For further investigation of the effects of hydrogen combustion on the EAF process, a pilot scale burner is operated inside a pilot scale EAF. Especially the effect on the melt and slag composition are analysed. These findings will help with the adjustment and validation of the hydrogen combustion in the EAF process model. Before the installation of the pilot scale burner in the EAF, free flame trials with fuel gas compositions ranging from 100% natural gas to 100% hydrogen were carried out to analyse the flame.

3. Results and discussion

In the following the results of different fuel gas compositions of the burners in the EAF process are shown. For a better comparison three different fuel gas compositions are analysed. The investigated compositions are as a reference 100% natural gas (NG), also a mixture of 50 vol-% natural gas and 50 vol-% hydrogen (NG_H2) and 100% hydrogen (H2). The axes of all shown graphs are normalised using the maximum of the investigated property. Furthermore, the process duration is also normalised. Each property is shown for two different EAFs. The two EAFs differ in their usage of burners. EAF 1 has more than double the amount of burner power than EAF 2. For this reason, the influence of the substitution of natural gas by hydrogen is more visible in the results from EAF 1.

In Figure 2 the normalised volume flow of non-combusted methane in the off-gas is shown for the different fuel gas compositions as well as the two different EAFs as they are simulated by the EAF process model. For a fuel gas composition of 100% hydrogen no methane is present in the off-gas, as it is not introduced by the natural gas through the burners. According to that for the mixture of natural gas and hydrogen less methane is in the off-gas than for 100% natural gas. Even though the graphs are normalised one can see that the burners are used more in EAF 1. In EAF 2 the burners are evidently only operated after a scrap basket is charged into the furnace, as a secondary heat source. In between the burners are completely turned off, as no methane is detected in the off-gas. In EAF 1 the burners are constantly operated with their power level adjusted over the duration of the heat.

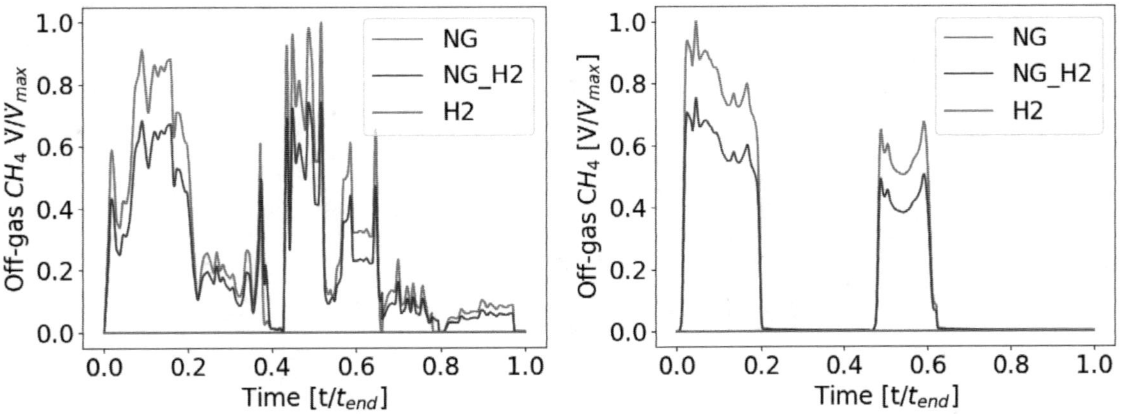

(a) EAF 1 (b) EAF 2

Figure 2. Methane content of the off-gas during one heat

Alongside that Figure 3 shows the simulated hydrogen content of the off-gas during an exemplary heat. In contrast to Figure 2 where the methane content for 100% hydrogen in the fuel gas is equal to zero, hydrogen is present in the off-gas for all analysed fuel gas compositions. Due to the high temperatures in the electric arc as well as in the furnace atmosphere, diatomic hydrogen gas is formed within the gas phase of the EAF. Therefore, it is also present in the off-gas. One can see that for both EAFs the hydrogen levels in the off-gas are increasing with rising hydrogen levels in the fuel gas. This difference indicates that unburned hydrogen introduced through the burners is present in the off-gas. In EAF 2 the difference is quite small and almost neglectable, as the burners are scarcely operated. For EAF 1 the hydrogen content in the off-gas is rising with increasing hydrogen levels in the fuel gas. At maximum the hydrogen level in the off-gas is doubled between the case of 100% natural gas and 100% hydrogen. However, one has to keep in mind the normalised scale. As the hydrogen content in the off-gas for 100% natural gas is fairly small, the resulting amount for 100% hydrogen is still quite small.

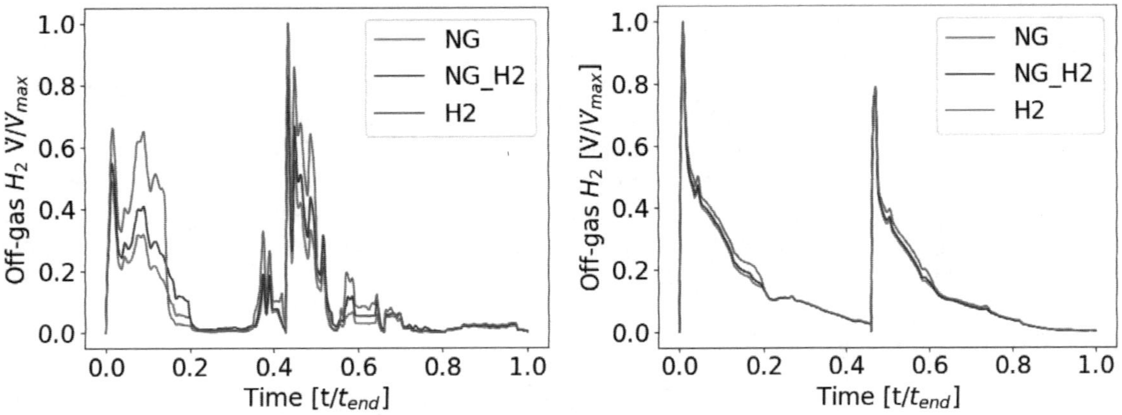

(a) EAF 1 (b) EAF 2

Figure 3. Hydrogen content of the off-gas during one heat

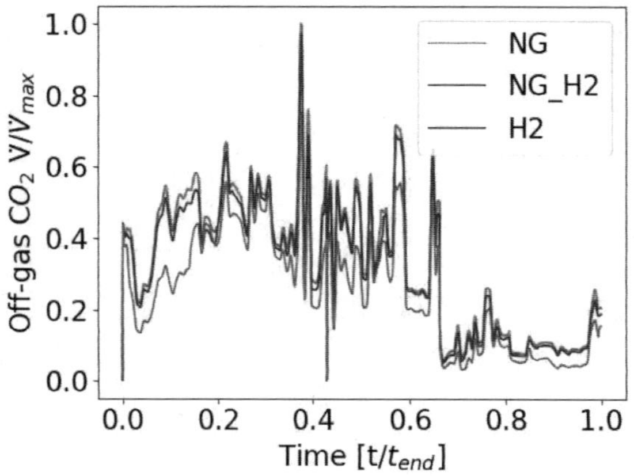

(a) EAF 1

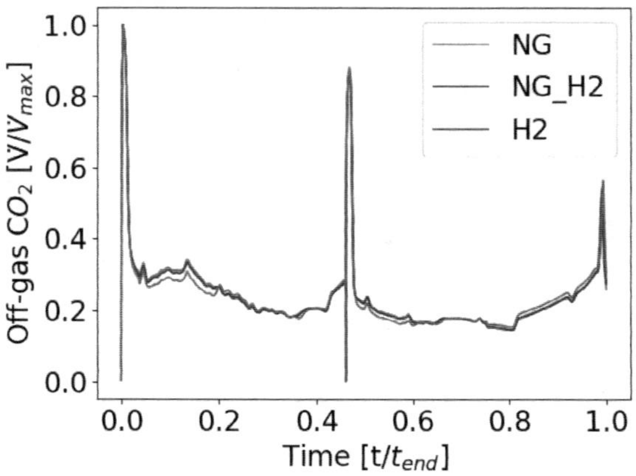

(b) EAF 2

Figure 4. Carbon dioxide content of the off-gas during one heat

The overall goal of the substitution of natural gas by hydrogen is the reduction of greenhouse gases especially carbon dioxide. Figure 4 shows the simulated carbon dioxide content in the off-gas for both EAFs. EAF 1 shows a significant carbon dioxide reduction in the off-gas with decreasing level of natural gas usage. For the fuel gas mixture of 50 vol-% natural gas and hydrogen the decrease in carbon dioxide emissions is small in comparison to 100% hydrogen. At maximum the carbon dioxide level of the off-gas is three times lower for 100% hydrogen than it is for 100% natural gas. According to this a reduction of up to 10% carbon dioxide is possible by substitution of natural gas with hydrogen. In Figure 4 for EAF 2 it is visible that this is only possible if the burners are frequently operated. For EAF 2 a reduction in carbon dioxide emissions is slightly visible at times of burner operation (e.g. Figure 2). However, almost no difference is apparent in the carbon dioxide emissions.

The change in the composition of the gas zone due to the change in fuel gas composition, also influences the radiative properties of the gas zone. In Figure 5 the simulated net radiation to and from the gas zone to other zones of the EAF process model is shown on a normalised scale. This radiation is calculated in megawatt in the process model. In EAF 2 no difference is visible between the different fuel gas compositions as the burners are used too infrequently to have an evitable effect on the radiative properties of the off-gas. With rising hydrogen levels in EAF 1 the radiative properties of the off-gas vary. The radiative properties of carbon dioxide are responsible for the climate damaging effects of carbon dioxide in the atmosphere. These properties are enhancing the radiation of the gas zone in the EAF. For this reason, the radiative properties of the off-gas are decreasing with rising hydrogen levels in the fuel gas. In the model the dust loading of the gas zone is not considered, therefore the effect of the hydrogen burners may be overestimated.

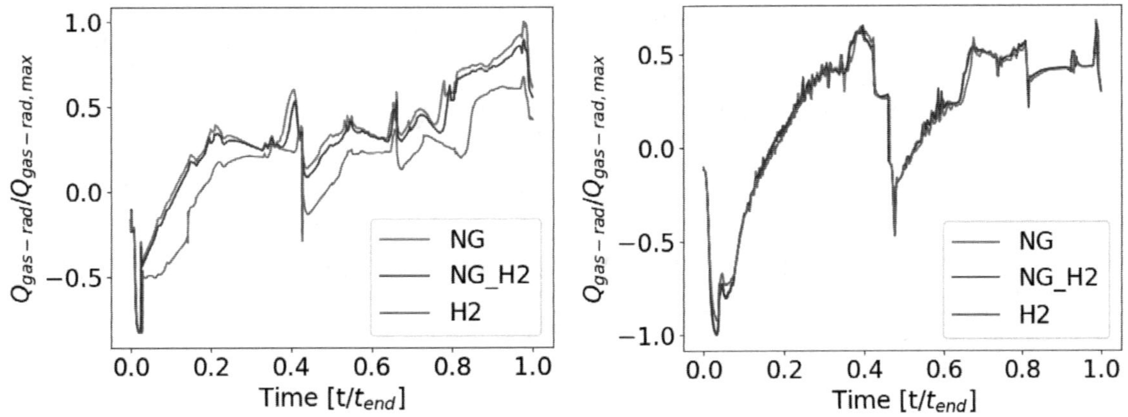

(a) EAF 1 (b) EAF 2

Figure 5. Radiative properties of the off-gas during one heat

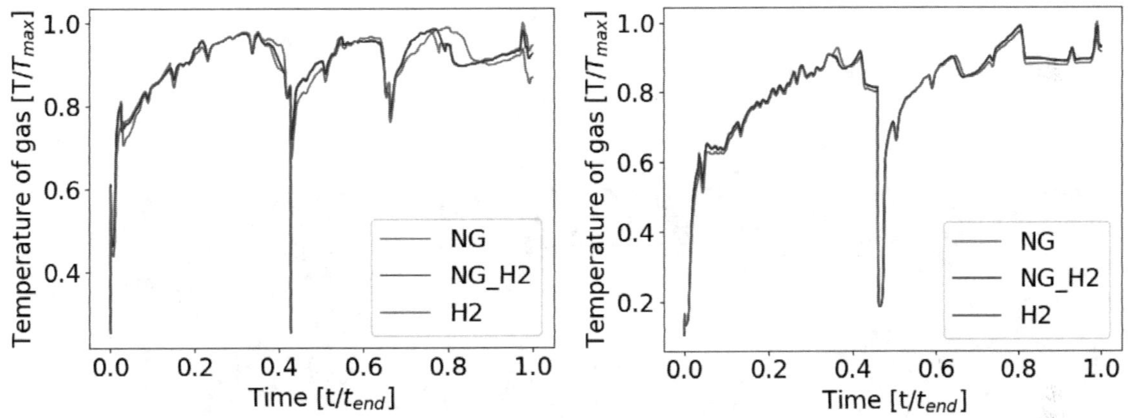

(a) EAF 1 (b) EAF 2

Figure 6. Temperature of the off-gas during one heat

Figure 6 shows the normalised temperature of the off-gas over the course of one heat for the different fuel gas compositions. As one can see the temperature of the off-gas does not change in the EAF process model with different fuel gas compositions for either furnace. This leads to the assumption that the change in radiative properties does not change the overall temperature in the gas zone.

Most important for the EAF operation is the influence of the substitution of natural gas by hydrogen on the temperature of liquid steel, as this temperature dictates the time for tapping. The normalised temperature of liquid steel is shown in Figure 7. For both EAFs the same influence of rising hydrogen levels in the fuel gas is visible, for EAF 1 the effects are better visible as the burners are used more often than in EAF 2. After the first phase where the burners are used, the temperature of liquid steel is higher for higher levels of hydrogen in the fuel gas. This may be accounted to the higher radiative properties of the off-gas with a hydrogen combustion. At tapping this trend changes and the temperature of liquid steel decreases with increasing hydrogen levels. This effect may be referred to the calculations in the model. Within the model, higher temperatures in the beginning of the heat result in an increased total energy loss through the water-cooling system and thus are responsible for the lower temperatures at tapping. Especially for the operation with 100% hydrogen the temperature of liquid steel is higher during the process, resulting in a bigger difference of tapping temperature than for 50 vol-% hydrogen. The cooling system is operated equally for all fuel gas compositions as real furnace data was taken as a reference for all in- and outgoing mass flows except for the fuel gas. In trials using hydrogen as a fuel gas the cooling system would be adjusted to fit the operation.

In the EAF operation these small differences due to the changes in the fuel gas composition can be cancelled out by adjustments in the operation chart. Furthermore, the shown results featuring a combustion of hydrogen are not validated yet. This validation will be carried out as soon as industrial trials at the investigated EAFs are carried out. An adjustment of the model parameters for the simulation of the hydrogen combustion may be possible.

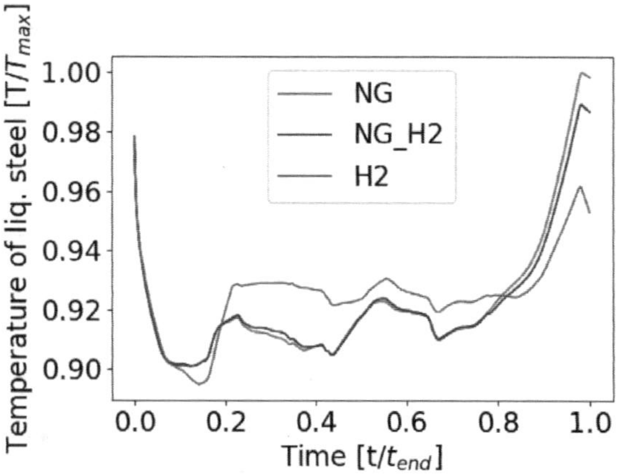

(a) EAF 1

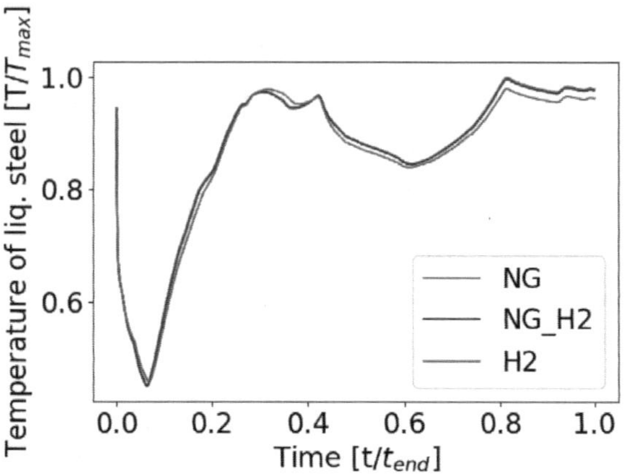

(b) EAF 2

Figure 7. Temperature of liquid steel during one heat

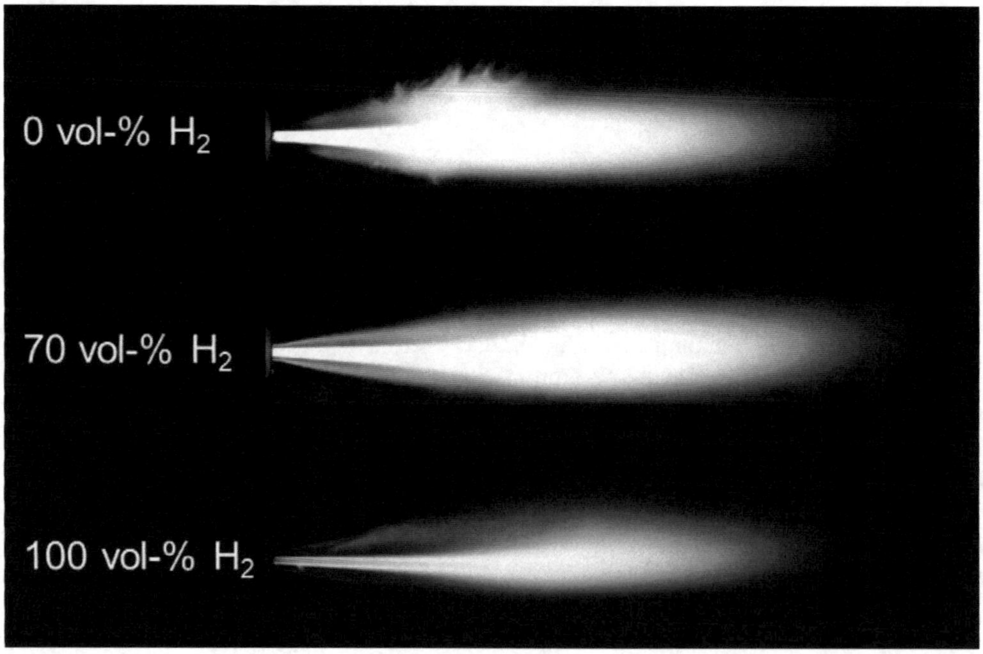

Figure 8. Free flame trials of 50 kW burner

Before the installation of hydrogen fired burners in the investigated EAFs a pilot scale burner will be installed in a pilot scale EAF and will be fired with fuel gas compositions ranging from 100% natural gas to 100% hydrogen. The burner is a scaled-down 50 kW burner-injector system which is especially designed for the combustion of natural gas and hydrogen. Through the inner laval nozzle a supersonic oxygen jet is formed, while the fuel gas is introduced through an annular gap. An analysis of the flame is carried out by free flame trials of this burner with different fuel gas compositions. These trials were carried out in ambient atmosphere and the caloric value was held constant at 50 kW with an air-fuel ratio of λ = 1.1. The results are shown in Figure 8. It is visible that for all fuel gas compositions a long and stable flame is produced. As an estimation, the visible flame for the natural gas combustion is about 1 meter in length. With the addition of hydrogen to the fuel gas the flame increasingly stabilised itself. For all fuel gas compositions, the length of the visible flame is in the same range. Even for the flame with 100% hydrogen in the fuel gas a flame is visible, due to particles present in the ambient atmosphere.

4. Conclusion

The described EAF process model is able to simulate the processes inside the EAF. This model is adjusted to be able to investigate the conditions within the furnace induced by the combustion of hydrogen. Two exemplary heats of different EAFs and with different burner usages were investigated for fuel gas compositions ranging from 100% natural gas to 100% hydrogen. The results show that hydrogen fired burners influence the overall process of the EAF operation. These are no major effects and can be cancelled out by small adjustments to the operation chart of the furnace. Nonetheless, the model has yet to be validated for hydrogen combustion by conducting industrial trials. The preliminary results show that with an intensive use of burners a carbon dioxide reduction up to 10% is possible by substitution of natural gas with hydrogen.

Before these industrial trials free flame trials of a pilot scale hydrogen burner were carried out to analyse the flame. With more hydrogen in the fuel gas the flame becomes more stable. Also, the flame length for all fuel gas compositions is sufficient for the application in the EAF. In a next step this pilot scale burner will be installed in a pilot scale EAF and the effects of a hydrogen fired burner on the EAF operation can be further analysed.

Acknowledgement

The research leading to these results has received funding from the European Community's Research Fund for Coal and Steel (RFCS) under the Grant Agreement number 101034081.

References

[1] World Steel Association 2022 *Steel Statistical Yearbook 2022* World Steel Association (Brussels Belgium)
[2] Steel manufacturers Association (SMA) 2022 *Emissions Analysis Executive Summary*
[3] Logar V and Škrjanc I 2012 Modeling and Validation of the Radiative Heat Tranfer in an Electric Arc Furnace *ISIJ Int.* **52/7** 1225-32
[4] Logar V and Škrjanc I 2012 Development of an Electric Arc Furnace Simulator Considering Thermal, Chemical and Electrical Aspects *ISIJ Int.* **52/10** 1924-26
[5] Logar V, Dovžan D and Škrjanc I 2012 Modeling and Validation of an Electric Arc Furnace Part 1, Heat and Mass Transfer *ISIJ Int.* **52/3** 402-12
[6] Logar V, Dovžan D and Škrjanc I 2012 Modeling and Validation of an Electric Arc Furnace Part 2, Thermo-chemistry *ISIJ Int.* **52/3** 413-23
[7] Logar V, Dovžan D and Škrjanc I 2011 Mathematical modeling and experimental validation of an electric arc furnace *ISIJ Int.* **51/3** 382-91
[8] Meier T 2016 *Modellierung und Simulation des Elektrolichtbogenofens (Dissertation)* (Aachen: Faculty of Georessources and Materials Engineering, RWTH Aachen University, Germany)
[9] Hay T, Hernandez J D and Roberts S 2021 Cacluation of View Factors in EAF Process Modeling *Steel Research Int.* **92/2**
[10] Hay T, Reimann A and Echterhof T 2019 Improving the Modeling of Slag and Steel Bath Chemistry in an Electric Arc Furnace Process Model *Metallurgical and Materials Transactions B* **50/5** 2377-88
[11] Hay T, Echterhof T and Visuri V-V 2019 Development of an Electric Arc Furnace Simulator Based on a Comprehensive Dynamic Process Model *Processes* **7/11** 852
[12] Hay T 2021 Mathematische *Modellierung des Elektrostahlverfahrens (Dissertation)* (Aachen: Faculty of Georessources and Materials Engineering, RWTH Aachen University, Germany)

[13] Hay T, Visuri V-V, Aula M, and Echterhof T 2020 A Review of Mathematical Process Models for the Electric Arc Furnace Process *Steel Research Int.* **92/3**

[14] MacRosty R D M and Swartz C L E 2005 Dynamic Modeling of an Industrial Electric Arc Furnace *Ind. Eng. Chem. Res.* **44**

[15] MacRosty R D M and Swartz C L E 2007 Dynamic Optimization of Electric Arc Furnace Operation *AlChE Jour.* **53/3**

[16] Rathaba L P 2004 *Model Fitting for Electric Arc Furnace Refining (Master Thesis)* (Pretoria: Faculty of Engineering, the Built Environment and Information Technology, University of Pretoria, South Africa)

IOP Conf. Series: Materials Science and Engineering 1309 (2024) 012012 doi:10.1088/1757-899X/1309/1/012012

Development of a new laboratory-scale reduction facility for the hydrogen plasma smelting reduction of iron ore based on a multi-electrode arc furnace concept

F Hoffelner[1], M A Zarl[2] and J Schenk[1,2]

[1] Lehrstuhl Eisen- und Stahlmetallurgie, Montanuniversität Leoben, Franz-Josef-Straße 18, Leoben 8700, Austria
[2] K1-MET GmbH, Stahlstraße 14, Linz 4020, Austria

felix.hoffelner@unileoben.ac.at

Abstract. Steel production accounts for a significant share of industrial CO_2 emissions. The HPSR process is a possible alternative to reduce these emissions massively if not completely negate them. In principle, Fe-ore is reduced at high temperatures in the plasma of a DC electric arc. hydrogen reacts with the oxidic melt at the gas-liquid interface. Various concepts for the hydrogen plasma reduction of iron ore have been investigated, but the process technology has not yet surpassed the demonstration scale (TRL5). Experimental setups for charging masses from a few grams to a few hundred kilograms have been realized. Further investigations on the process stability and the reaction kinetics are still necessary. An improved laboratory-scale furnace concept shall provide the basis for the fundamental research. An existing laboratory facility is the starting point for designing and constructing the new plasma furnace. There are several problems with this experimental setup. Mainly, the reactor's dimensions and power supply limitations restrict the arc's length. The first leads to problems with excessive refractory wear, while the latter limits the variation of process parameters. Strong cooling when using Fe crucibles and the unstable nature of the arc complicate the process control. A promising concept to deal with the problem of arc stability is the use of multiple electrodes in a direct current arc furnace. Together with an optimized furnace geometry, new potential for further investigations can open. Using a multi-cathode furnace is also promising to further explore ferroalloy production via hydrogen plasma reduction. An electric arc furnace was designed based on the requirements for the planned plasma reduction facility. The energy requirement was based on assumptions for heat transfer from the arc to the melt, walls, and lid and continuous transfer through the individual furnace parts. Considerations of power supply, hearth dimensions, refractory design, controlled gas atmosphere, and the implementation of auxiliary equipment were central to creating an ideal basis for various experimental setups.

1. Introduction

Steel production accounts for 7% of worldwide greenhouse gas emissions [1]. As the predominant production route, we know today, the blast furnace process results from historical development and optimization. The necessity of using coke as a reducing agent limits the potential for further reductions in CO_2 emissions to a minimal degree. Additionally, the inherent nature of the process prevents the complete mitigation of greenhouse gas emissions. hydrogen plasma smelting reduction (HPSR) is a possible alternative to reduce these emissions massively if not completely negate them [2]. The process reduces Fe-oxides from a liquid phase of molten iron ore in a single reactor. This is one crucial benefit compared to the blast furnace process, which must be coupled with a basic oxygen furnace to produce crude steel. In principle, Fe-ore is reduced at high temperatures in a transferred direct current (DC) electric arc plasma. The current hypothesis is that hydrogen reacts with the oxidic melt at the gas-liquid interface in the direct vicinity of the electric arc via equations (1)–(6) (species are denoted (), {}, [] according to their presence in the slag, gas, and metallic phases) [3,4]. The according reaction enthalpies are displayed at 1600 °C. This temperature was chosen as a requirement for the presence of a continuous liquid metal bath in the presented furnace concept. From the molecular H_2, introduced as the reducing agent, atomic H and ionized H^+ are generated in the plasma of the electric arc. While all species are assumed to contribute to the overall reaction, H^+ is only present in limited amounts.

Extremely high temperatures prevail in the plasma region. Temperatures of up to 10000 K were derived from Optical Emission Spectroscopy (OES) measurements under process conditions [5]. Calculations with FactSage 8.2 predict that a mixture of atomic H with approximately 2.7% H^+ would be present under equilibrium conditions at these temperatures [3]. Collecting detailed information on the contribution of each species to the overall reaction seems challenging, and the authors have not come across such information to date. Although considering ionization processes in the plasma column, a distinct possibility for (5) and (6) cannot be denied.

$$2(Fe^{3+}) + (O^{2-}) + \{H_2\} \rightarrow 2(Fe^{2+}) + \{H_2O\} \qquad \Delta H_r^{1600°C} = 24.3\ kJ \qquad (1)$$

$$2(Fe^{3+}) + (O^{2-}) + \{H_2\} \rightarrow 2(Fe^{2+}) + \{H_2O\} \qquad \Delta H_r^{1600°C} = 24.3\ kJ \qquad (2)$$

$$2(Fe^{3+}) + (O^{2-}) + 2\{H\} \rightarrow 2(Fe^{2+}) + \{H_2O\} \qquad \Delta H_r^{1600°C} = -429\ kJ \qquad (3)$$

$$(Fe^{2+}) + (O^{2-}) + 2\{H\} \rightarrow [Fe] + \{H_2O\} \qquad \Delta H_r^{1600°C} = -462\ kJ \qquad (4)$$

$$2(Fe^{3+}) + (O^{2-}) + 2\{H^+\} + 2e^- \rightarrow 2(Fe^{2+}) + \{H_2O\} \qquad \Delta H_r^{1600°C} = -2.95\ MJ \qquad (5)$$

$$(Fe^{2+}) + (O^{2-}) + 2\{H^+\} + 2e^- \rightarrow [Fe] + \{H_2O\} \qquad \Delta H_r^{1600°C} = -2.98\ MJ \qquad (6)$$

To implement the HPSR process on a larger scale with sufficient productivity, precise knowledge of the overall reaction and its kinetics is necessary. Although a detailed understanding of the mechanistic processes would be preferable, an accurate representation of the overall reduction progress and its dependencies is sufficient for practical purposes. In addition to the kinetics of the HPSR process, it is also essential to investigate the required conditions to maintain a stable arc in atmospheres with a high H_2 content in the gas feed. An adapted furnace concept needs to be developed to address these challenges. Several concepts for the hydrogen plasma reduction of iron ore have been investigated, such as liquid film reactors with non-transferred arc, in-bath reduction facilities with transferred arc, and in-flight reduction with fine ore injected into the arc zone [6]. However, HPSR technology has not yet surpassed the demonstration scale. Experimental setups with charge masses ranging from a few grams to a few hundred kilograms have been realized.

Investigations on thermodynamically more stable oxides such as Cr_2O_3 or TiO_2 have shown that these also can be reduced with hydrogen plasmas. Kitamura et al. [7] reported that Cr_2O_3 could be reduced to the metallic state via in-flight reduction. TiO_2 and Al_2O_3 were shown to be vaporized as their suboxides and then precipitate as stable oxides. Plaul [8] conducted reduction trials on Fe-ore by adding Cr_2O_3 with an HPSR setup. The results indicated that reducing a Fe-Cr alloy was possible, but higher energy input would be necessary for high Cr yields. The new HPSR laboratory facility could be used to investigate the potential for ferroalloy production via this process.

Several groups studied the HPSR process using different setups over the past 40 years. Nakamura et al. [9] used a single W cathode on iron ore in a water-cooled crucible. 500 g of ore were reduced with arc powers of 15 kW using Ar-H_2-mixtures with 10–50% H_2. Voltages ranged from 40–

60 V with currents of 250–350 A. Kamiya et al. [10] also performed reduction experiments with transferred DC arcs on FeO-containing oxide melts. They used cooled Cu crucibles and MgO refractory crucibles in different experimental setups. Experiments with 25–75 g pure Fe_2O_3 were conducted in the cooled crucible. Experiments in MgO crucibles were performed on pre-reduced fine ore dissolved in synthetic slag above liquid Fe. Behera et al. [11] performed experiments in a larger setup with charge masses of up to 7 kg. A water-cooled Cu crucible was used in conjunction with a non-transferred plasma torch using a W cathode and a Cu anode. For charge masses of 1000 g, a power of 12–15 kW was specified. They assessed a variety of bath depths with crucible diameters of 100 and 200 mm. Their findings suggested that the depth-diameter ratio of the oxide bath limits the possible reduction degree. Too shallow baths do not allow full Fe recovery, according to their results. Souza Filho et al. [12] experimented with a transferred arc in a cooled Cu crucible. Their procedure consisted of interrupted reduction cycles of 1 min each. During this time, they used an arc with a relatively high power of 35 kW on only a few grams of charged Fe_2O_3 material. Between the reduction periods, the gas atmosphere was renewed to an Ar-10% H_2 mixture while the reactor was sealed entirely during the reduction time. Bäck, Plaul, Zarl, et al. [2,13–15] extensively researched the HPSR process with a transferred arc system. The setup consisted of a single graphite cathode with Fe-crucibles with a power input from 7–15 kW. Samples of approximately 100 g ore were used to investigate various aspects of the process.

Crucibles of laboratory-scale arc furnaces for the HPSR process are mostly heavily cooled. While this approach makes sense for very small facilities and certain experiments, it hinders or, to some extent, prevents other experimental setups. To further investigate the HPSR process, a furnace design with a constant pool of molten metal and oxide is of interest. This enables experimental conditions closer to those of the envisioned industrial HPSR process. Typically, laboratory-scale DC EAFs are designed as insulated systems with limited cooling. The only parts to be extensively cooled are design aspects that require increased heat fluxes, such as a bottom anode, measuring equipment or refractory concepts. Table 1 presents data on the power requirements of several laboratory-scale furnaces, which will be compared against the results obtained in this study.

Table 1. Data on operational ranges for power supplies of laboratory-scale EAFs.

Type	Capacity (kg)	Power (kW)	Voltage (V)	Current (A)	Reference
DC	100	350	30–250	2000	HPSR Donawitz
DC	45	20	80	400	Knight et al. [16]
DC	12	50	35–50	1250	Xue et al. [17]
AC with a bottom electrode	3	25	85	300	Mandal et al. [18,19]

A demonstrational reduction plant for the HPSR process with a charge weight of around 100 kg is operated by K1-MET at voestalpine Stahl Donawitz in Leoben. This facility represents a scaled-up version of the existing laboratory facility at the Montanuniversität Leoben. The new furnace shall bridge the gap between existing 100 g and 100 kg facilities for in-bath reduction. An additional motivation for constructing a multi-cathode furnace (MCF) is to expand the possibilities for investigations on the HPSR process compared to the existing facility. Primarily investigations on the kinetics are interesting to better understand the reactions involved. Also, the possibility of introducing the reactants directly into the hotspot of the plasma gives rise to new possibilities regarding process control and stability.

2. Design and calculations

The starting point for designing and constructing the new plasma furnace is the laboratory facility with about 100 g of feed material. This design is intended to melt pre-charged ore and a continuous feed of ore fines during the reduction. H_2 is injected via a hollow electrode, reducing the molten iron ore. An iron crucible or a pin-type electrode inside a refractory crucible is the second electrode to close the electric circuit. While the existing facility allows refractory crucibles, it was initially designed to use steel crucibles. Experiences with the demonstrational plant in Leoben also influenced the design decisions. The proposed reactor design also consider relevant aspects based on studies of the HPSR process and DC EAFs, which are discussed in the introduction section.

Several problems exist in the experimental setup of the laboratory facility. The dimensions of the system and limitations regarding the supplied voltage restrict the length of the arc. This also leads to problems with premature wear of crucibles made from refractory material. During operation in an iron crucible, strong cooling complicates the process control. The unstable nature of the arc is also problematic, which makes process control even more difficult. The mentioned limitations restrict investigations on larger parameter areas in the existing furnace.

To cover the planned investigations, the scope of the furnace must meet some essential requirements. Multiple electrodes in parallel will be used to investigate their influence on the stability of the arc column. The objective is to create one central hotspot between the single independently powered electrodes, as the single arc columns are expected to attract each other [20]. This opens opportunities for experiments with varying arc power and hotspot size and for using different numbers of electrodes. It also presents the possibility of introducing charge material of various constitutions directly into the hot spot. If a more stable arc can be achieved in the new setup, OES measurement in the hot spot would allow further investigations on the plasma conditions. The new concept will represent the basis for further optimizations of the HPRS process. As the furnace is to be used in a laboratory environment with various parameters, the focus is on achieving the highest possible flexibility rather than efficient operation. Therefore, the furnace design aspects and associated calculations are liberal estimates and not fully optimized results.

The thermal losses for the proposed concept were estimated to determine the necessary power for the furnace. These calculations are based on a rudimentary replacement system, which can be seen in Figure 1. The overall transfer efficiency from an electric arc to an open bath was determined by Ameling et al. [21] to be 36%. Due to the presence of an H_2 atmosphere, it was assumed that a shorter arc was present, leading to a value of 40% being used during calculations. In their research, Alexis et al. [22] studied heat transfer from the arc to the molten bath through different mechanisms. Although using higher power in the MW range for their calculations, they found that shorter arcs with lower currents are more effective for energy transfer to the bath. They determined that radiative transfer, convective transfer, and heat generated by anodic effects contributed 42%, 36%, and 22%, respectively. This work utilized these findings to make estimations and calculate the radiative heat transfer from the arc to the bath. The calculation of heat transfer between the solid surfaces within the furnace and the bath in this study only accounted for radiative exchange and assumed that the bath, wall, lid, and cathodes had isothermal surfaces.

The view factors between the surfaces were calculated with Python. A rudimentary model of the inner furnace geometry was created with the PyVista module, and the view factors were calculated based on the PyViewFactor module [23,24]. The view factors were calculated with three cathodes of 16 mm diameter in place, including their shadowing effect. The calculated view factors exhibit good agreement with those calculated by Hay et al. [25] for an open bath configuration. Conductive heat transfer through the refractory materials and the steel shell was calculated from the inner surfaces. Although we are dealing with a small geometry with comparatively thick refractory layers, the transfer was assumed to be one-dimensional trough planes (8) and cylinder (9) areas, and additional terms for the edges were added in parallel [26]. Conduction through the anode and the cathode were also included in the calculation. The thermal resistances of different furnace layers and contact resistances were added in series and parallel to respective replacement values for the hearth, the wall, and the lid. The heat flux through these three parts was then calculated according to (7) [26]. Heat transfer from the outer steel shell was assumed as radiative and convective transport against surroundings at 30 °C. Wall and hearth thicknesses for the insulation layer were adjusted so that the outer surfaces would not surpass 200 °C.

The outer surface of the lid is additionally considered to be actively cooled to a temperature of 70 °C. This is especially important as it accommodates the cathodes and auxiliary equipment. Furthermore, it must maintain a secure seal and prevent water condensation. The necessary power input from the arc was calculated on the premise that temperatures of 1850 K for the bath could be assured. Figure 1 shows the cylindrical replacement system for the performed calculations, and the data used for the calculations can be found in Table 2. Following are the equations used for the calculation of the thermal fluxes:

$$q = \frac{\Delta T}{\sum R_{\text{th}}} \tag{7}$$

$$R_{\text{th,plane}} = \frac{\Delta x}{k \cdot A} \tag{8}$$

$$R_{th,cylinder} = \frac{ln\left(\frac{r_o}{r_i}\right)}{2\pi k l} \tag{9}$$

where q is the thermal flux (W), R_{th} the thermal resistance ($W \cdot K^{-1}$), ΔT the temperature difference (K), Δx the wall thickness (m), k the thermal conductivity ($W \cdot m^{-1} \cdot K^{-1}$), A the area of the plane (m^2), r_o and r_i the outer and inner radius (m), and l cylinder length (m).

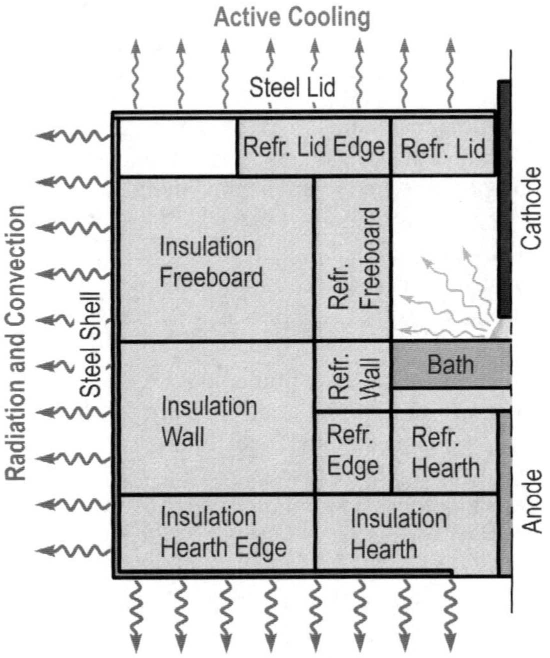

Figure 1. Geometry of the model assumptions for calculating thermal losses.

H_2 and a stabilizing gas will be used during the process. While a process with pure H_2 would be desirable, a certain amount of stabilizing gas may be necessary even at higher voltages. The enthalpies to heat the process gases are included in the thermal balance to calculate the power requirements. The necessary electrical power is calculated for various operating conditions in the steady state. The goal is to guarantee a stable operation throughout all phases of reduction trials. This includes furnace heat-up, batch-wise operation, operation with a continuous charge feed, and tapping the liquid metal.

Additional to the thermal balance, various design aspects for the MCF were considered. These include placement and control of the electrodes, placement for gas injection, and design of the furnace vessel and lid to house all auxiliary equipment.

Table 2. Parameters used for the calculations of thermal losses.

Parameter	Value	Unit	Ref.
Thermal conductivity refractory crucible and lid	5	$(W \cdot m^{-1} \cdot K^{-1})$	[27]
Thermal conductivity insulating bricks	0.6	$(W \cdot m^{-1} \cdot K^{-1})$	[27]
Thermal conductivity graphite	160	$(W \cdot m^{-1} \cdot K^{-1})$	[27]
Thermal contact resistance steel-refractory	0.0036	$(m^2 \cdot K \cdot W^{-1})$	[28]
Convective transfer coefficient hearth	2.4	$(W \cdot m^{-2} \cdot K^{-1})$	[29]
Convective transfer coefficient wall	5	$(W \cdot m^{-2} \cdot K^{-1})$	[29]
Convective transfer coefficient lid	6	$(W \cdot m^{-2} \cdot K^{-1})$	[29]
Emissivity refractory	0.7		[30]
Emissivity oxide melt	0.8		[31]
Emissivity coated furnace shell	0.91		[32]
Specific heat refractory magnesia aluminate T_{avg}= 1300 °C	1400	$(J \cdot kg^{-1} \cdot K^{-1})$	[3]
Specific heat isolation calcium silicate T_{avg}= 640 °C	1050	$(J \cdot kg^{-1} \cdot K^{-1})$	[3]
Specific heat steel shell T_{avg}= 150 °C	500	$(J \cdot kg^{-1} \cdot K^{-1})$	[3]

3. Resulting design aspects and discussion

To perform extensive experiments on the HPSR process on a laboratory scale, the most important goal is to design the laboratory facility as close to the envisioned industrial process as possible. This includes the constant presence of a molten Fe-bath during the in-bath reduction process. Each electrode in the reactor requires a minimum distance to the side walls. Multiple electrodes must also be separated far enough to obtain a usable area at the hotspot. The resulting large bath area leads to a relatively shallow pool. Limiting is mainly the charge weight to restrict the size, complexity, and expenses of the laboratory trials.

Nevertheless, it is necessary to maintain a minimum bath depth to ensure a constant pool across the entire furnace area and prevent the arc from directly burning onto the anode and its surrounding refractories. The consequence of these considerations leads to the necessity of a constant Fe bath of 1 kg below the oxidic melt. The MCF is intended to be used with 2–7 kg Fe-ore charge masses.

The goal to maintain a constant molten Fe bath beneath the molten oxide layer also opens the possibility of tapping the contents of the reactor. Therefore, a tilting vessel with a spout is planned. This opening also functions as the off-gas channel with the benefit of preheating it for easier handling during tapping. The anode at the bottom of the vessel will consist of several sheet metal fins made from pure iron strips surrounded by a ramming mass. The refractory concept consists of an insulating layer of lightweight refractory bricks and a cast wear lining in contact with the melt. As the furnace is designed for laboratory-scale experimentation, adaptations can easily be made if required. Gas will be supplied by an array of mass flow controllers (MFC). Three pairs of MFCs will distribute H_2 and a stabilizing gas to the single electrodes. Alternatively, they will be used together if the gas is supplied directly to the hot spot.

Assuming a heat-up time for the charge of 30 min and a reduction time of 120 min, the power requirements for the various operational states were calculated. Calculations for the thermal balance result in a required furnace power of 11 kW to maintain a molten bath in the steady state, which constitutes mainly losses. During reduction operation, the primary energy sinks are the reduction of Fe^{3+} to Fe^{2+} and the required enthalpy to heat the process gases and the charged material. For the different operational states, the required power results in 21, 15, and 18 kW for batch-wise heating, batch-wise reduction, and reduction under continuous feed. These calculations do not consider the necessary energy to heat the furnace to steady-state temperatures. As this would take at least 260 min at a furnace power of 21 kW, it seems necessary to pre-heat the refractory material to a certain degree. The most sensible solution is to use a resistive heater, as the power supply for the arc is already present. The results for the furnace power are in good agreement with the literature data on laboratory-scale EAFs, and a power source with a maximum power of 50 kW seems reasonable for the task at hand. Voltages up to 150 V and current sums up to 350 A are planned for the upper limits.

Three galvanically isolated DC power sources are required because three independent arcs are operated simultaneously on the bath as the anode. Compared to other plasma gases, the voltage drop over an arc length is high in an H_2 atmosphere. This restricts the lower limit voltage, as too low values can lead to arc detachment due to the fluctuating nature of a transferred plasma. At the same time, the power supply must provide a wide range of voltages, currents, and power levels. As one of the main objectives of furnace design is to study kinetic effects and arc stability, the broadest possible variation of these parameters is required.

Keeping a gas-tight seal between all furnace parts is vital for safety reasons. While sealing the off-gas duct and the lid to the furnace vessel is relatively simple. The main design challenge is the lid assembly with the retractable cathodes and additional openings for gas supply, sampling, and observation. The relatively tight space must fit all the equipment while still thermally insulating the furnace. Additional ports in the lid and vessel are planned for sampling and as a provision for future measuring device implementation. Figure 2 shows the preliminary schematics for the furnace design. The three electrodes are positioned circularly around an additional lance for gas supply and injection of fines into the center.

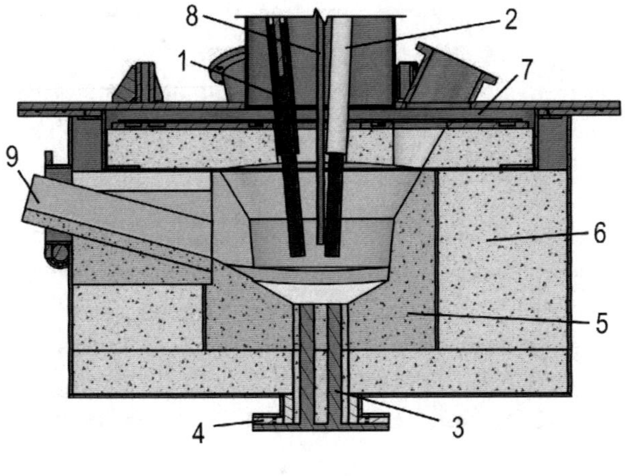

1 Cathode 4 Isolation 7 Cooled lid
2 Isolation 5 Refractory 8 Injection lance
3 Anode 6 Insulation 9 Spout

Figure 2. Preliminary schematics for the MCF vessel and refractory design.

4. Conclusion

An MCF DC EAF presents a promising concept for further investigations on the HPSR process. The key requirement is a design as close as possible to the envisioned industrial process. As a result, the following aspects are held to be crucial for the intended task:

- Multiple separately supplied electrodes
- Refractory lined and insulated furnace vessel to create near-process conditions and allow tapping of the contents
- Cooled lid construction to house different auxiliary devices and easy adaption

The engineering work for the laboratory-scale facility is expected to be completed by the third quarter of 2023, with construction targeted for completion by the third quarter of 2024. The data generated from this facility will be crucial in advancing the HPSR process toward technological readiness, particularly in understanding its kinetics.

References

[1] International Energy Agency (IEA) 2020 Iron and Steel Technology Roadmap: Towards more sustainable steelmaking *https://www.iea.org/reports/iron-and-steel-technology-roadmap* accessed 25 April 2023

[2] Plaul J F, Krieger W and Bäck E 2005 Reduction of Fine Ores in Argon-Hydrogen Plasma *steel research int.* **76** 548–54

[3] Bale C W *et al* 2016 FactSage thermochemical software and databases, 2010–2016 *Calphad* **54** 35–53

[4] Hiebler H and Plaul J F 2004 Hydrogen Plasma Smelting Reduction - an Option for Steelmaking in the Future *Metalurgija* **43** 155–62

[5] Pauna H, Ernst D, Zarl M, Aula M, Schenk J, Huttula M and Fabritius T 2022 Hydrogen plasma smelting reduction process monitoring with optical emission spectroscopy – Establishing the basis for the method *Journal of Cleaner Production* **372** 133755

[6] Sabat K C and Murphy A B 2017 Hydrogen Plasma Processing of Iron Ore *Metall and Materi Trans B* **48** 1561–94

[7] Kitamura T, Shibata K and Takeda K 1993 In-flight Reduction of Fe2O3, Cr2O3, TiO2 and Al2O3 by Ar-H2 and Ar-CH4 Plasma *ISIJ International* **33** 1150–8

[8] J.F. Plaul 2005 Schmelzreduktion von hämatitischen Feinerzen im Wasserstoff-Argon-Plasma *Doctoral Thesis* Leoben

[9] Nakamura Y, Ito M and Ishikawa H 1981 Reduction and dephosphorization of molten iron oxide with hydrogen-argon plasma *Plasma Chem Plasma Process* **1** 149–60

[10] Kamiya K, Kitahara N, Morianka I, Sakuraya K, Ozawa M and Tanaka M 1984 Reduction of Molten Iron Oxide and FeO Bearing Slags by H2-Ar Plasma *ISIJ Int.* **24** 7–16

[11] Behera P R, Bhoi B, Paramguru R K, Mukherjee P S and Mishra B K 2019 Hydrogen Plasma Smelting Reduction of Fe2O3 *Metall and Materi Trans B* **50** 262–70

[12] Souza Filho I R, Ma Y, Kulse M, Ponge D, Gault B, Springer H and Raabe D 2021 Sustainable steel through hydrogen plasma reduction of iron ore: Process, kinetics, microstructure, chemistry *Acta Materialia* **213** 116971

[13] E. Bäck 1998 Schmelzreduktion von Eisenoxiden mit Argon-Wasserstoff-Palsma *Doctoral Thesis* Leoben

[14] Naseri Seftejani M, Schenk J and Zarl M A 2019 Reduction of Haematite Using Hydrogen Thermal Plasma *Materials (Basel, Switzerland)* **12**

[15] Zarl M A, Farkas M A and Schenk J 2020 A Study on the Stability Fields of Arc Plasma in the HPSR Process *Metals* **10** 1394

[16] Knight R, Murawa M J, Girgis N M and Reide K J 1990 Arc characteristics in small-scale DC plasma arc furnaces using graphite cathodes *Plasma Chem Plasma Process* **10** 359–73

[17] Xue B, Yang L, Guo Y, Chen F, Wang S, Zheng F and Yang Z 2021 Design and Construction of a Laboratory-Scale Direct-Current Electric Arc Furnace for Metallurgical and High-Titanium Slag Smelting Studies *Metals* **11** 732

[18] Mandal A K, Dishwar R K and Sinha O P 2018 Behavior of an indigenously fabricated transferred arc plasma furnace for smelting studies *Plasma Sci. Technol.* **20** 35506

[19] Mandal A K and Sinha O P 2016 Recovery of Multi-Metallic Components from Bottom Ash by Smelting Reduction Under Plasma Environment *Metall and Materi Trans B* **47** 19–22

[20] Reynolds Q G and Jones R T 2006 Twin-electrode DC smelting furnaces—Theory and photographic testwork *Minerals Engineering* **19** 325–33

[21] Ameling D, Petry J, Sittard M, Ullric W and Jose W 1986 Untersuchungen zur Schaumschlackenbildung im Elektrolichtbogenofen *Stahl und Eisen* **106** 625-620
[22] Alexis J, Ramirez M, Trapaga G and Jönsson P 2000 Modeling of a DC Electric Arc Furnace. Heat Transfer from the Arc *ISIJ International* **40** 1089–97
[23] Sullivan C and Kaszynski A 2019 PyVista: 3D plotting and mesh analysis through a streamlined interface for the Visualization Toolkit (VTK) *JOSS* **4** 1450
[24] Bogdan M, Walther E, Alecian M and Chapon M Calcul des facteurs de forme entre polygones - Application à la thermique urbaine et aux études de confort in: 2022 *IBPSA France*
[25] Hay T, Hernandez J, Roberts S and Echterhof T 2021 Calculation of View Factors in Electric Arc Furnace Process Modeling *steel research international* **92** 2000341
[26] J.P. Holman 2010 Heat transfer 10th ed. Boston, Mass. McGraw-Hill Higher Education
[27] S. Biswas 2020 Introduction to Refractories for Iron- and Steelmaking Cham Springer International Publishing AG
[28] Jain I, Singh R K and Mazumdar D 2015 Measurements of Some Thermal Properties of Steel-Refractory Systems and Heat Losses from Steelmaking Furnaces *Trans Indian Inst Met* **68** 383–92
[29] Trejo E, Martell F, Micheloud O, Teng L, Llamas A and Montesinos-Castellanos A 2012 A novel estimation of electrical and cooling losses in electric arc furnaces *Energy* **42** 446–56
[30] A.T. Vázquez 2019 Implementation of a Thermo Energetic Model of an Electric Arc Furnace Using Static Approach *Master's Thesis* Monterrey
[31] Khodabandeh E, Ghaderi M, Afzalabadi A, Rouboa A and Salarifard A 2017 Parametric study of heat transfer in an electric arc furnace and cooling system *Applied Thermal Engineering* **123** 1190–200
[32] LabIR Datasheet Thermographic paint for high temperature: HERP-HT-MWIR-BK-11 *https://paints.labir.eu/homepage/thermographic-paint-for-high-temperature-applications* accessed 5 May 2023

AUTHOR INDEX